CHEMISTRY DEMYSTIFIED

Other Titles in the McGraw-Hill Demystified Series

Astronomy Demystified by Stan Gibilisco
Calculus Demystified by Steven G. Krantz
Physics Demystified by Stan Gibilisco

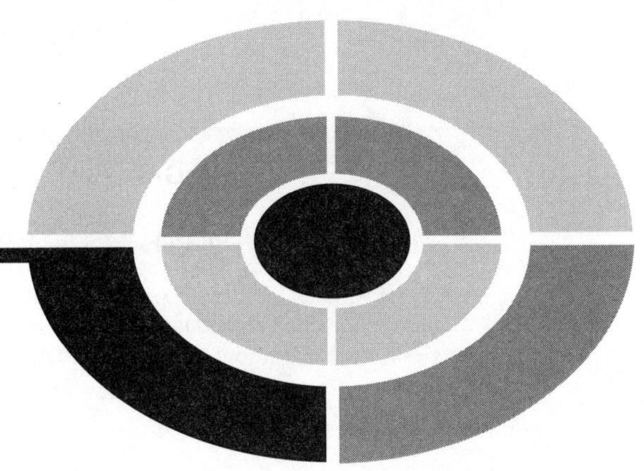

CHEMISTRY DEMYSTIFIED

LINDA D. WILLIAMS

McGRAW-HILL
New York Chicago San Francisco Lisbon London
Madrid Mexico City Milan New Delhi San Juan
Seoul Singapore Sydney Toronto

The McGraw-Hill Companies

Cataloging-in-Publication Data is on file with the Library of Congress.

9 0 DOC/DOC 0 9 8 7 6

ISBN 0-07-141011-2

The sponsoring editor for this book was Scott Grillo and the production supervisor was Sherri Souffrance. It was set in Times Roman by Keyword Publishing Services.

Printed and bound by RR Donnelley.

This book was printed on recycled, acid-free paper containing a minimum of 50% recycled, de-inked fiber.

Library of Congress Cataloging-in-Publication Data

Williams, Linda D.
Chemistry demystified/Linda D. Williams.
p. cm.
Includes bibliographical references and index.
ISBN 0-07-141054-6 (acid-free paper)
1. Chemistry. I. Title.

QD33.2.W54 2003
540–dc 21 2003052654

CONTENTS

PREFACE

This book is for anyone who has an interest in chemistry and wants to learn more about it outside of a formal classroom setting. It can also be used by home-schooled students, tutored students, and those people wishing to change careers. The material is presented in an easy-to-follow way and can be best understood when read from beginning to end. However, if you just want more information on specific topics like radioactivity or organic chemistry, then those sections can be reviewed individually as well.

You will notice through the course of this book that I have mentioned many milestone accomplishments of chemists, physicists, biochemists, and physicians. In particular, I have noted when a new discovery earned a Nobel prize for excellence and the advancement of science. I have highlighted these achievements to give you an idea of how much the questions and bright ideas of curious people (who just happen to like science) have brought to humankind.

Science is all about curiosity and the desire to find out how something happens. Nobel prize winners were once students who daydreamed about new ways of doing things. They knew answers had to be there and they were stubborn enough to dig for them. The Nobel prize for science has been awarded over 470 times. (Don't worry I haven't described every prize in this book.) However, to give you an idea of chemistry's diversity, I have listed some of the research areas that the Nobel (actors have Oscar and scientists have Nobel) has touched since 1901:

- isolation of fluorine
- fermentation and investigations in biological chemistry
- catalysis and investigations of chemical equilibrium and reaction rates
- discovery of the elements radium and polonium
- methods of hydrogenating organic compounds

- linking up atoms within the molecule
- investigations on dipole moments and diffraction of X-rays and electrons in gases
- isolating the coloring compounds of plants, especially chlorophyll
- discovery of the origin and nature of isotopes
- understanding atomic fission
- discovery of the molecular structure of insulin
- electronic structure and geometry of molecules, particularly free radicals
- deciphering the structure of biological molecules like antibiotics and cholesterol
- developing methods to map the structure and function of DNA
- discovering the detailed structures of viruses
- development of direct methods to determine crystal structures
- refinements in nuclear magnetic resonance spectroscopy
- understanding chemical processes that deplete the earth's ozone shield
- discovery of a new class of carbon molecule (fullerenes)
- invention of the world's fastest camera that captures atoms in motion.

In 1863, Alfred Nobel experienced a tragic loss in an experiment with nitroglycerine that destroyed two wings of the family mansion and killed his younger brother and four others. Nobel had discovered the most powerful weapon of that time, dynamite.

By the end of his life, Nobel had 355 patents for various inventions. After his death in 1896, Nobel's will described the establishment of a foundation to create five prizes of equal value "for those who, in the previous year, have contributed best towards the benefits for humankind" in the areas of chemistry, physics, physiology/medicine, literature, and peace. Nobel wanted to recognize the heroes of science and encourage others in their quest for knowledge. My hope is that in including some of the Nobel prize winners in this text you too will be encouraged by the success and inventiveness of earlier scientists who were curious to know how and why things happen.

This book provides a general overview of chemistry with sections on all the main areas you'll find in a chemistry classroom or individual study of the subject. The basics are covered to familiarize you with the terms and concepts most common in experimental sciences like chemistry. There is a Periodic Table printed on the inside cover of this book, as well as in Chapter 4 to use as a reference. Additionally, I have listed a couple of Internet sites on the Periodic Table that have a lot of good information. The Periodic Table is the single most useful tool in the study of chemistry beside the pencil. The complete description of the Periodic Table and its uses is described in Chapter 4.

Throughout the text I have provided examples for you to work, as well as quiz, test, and exam questions. All the questions are multiple choice and very much like those used in standardized tests. There is a short quiz at the end of each chapter. These quizzes are "open book." You shouldn't have much trouble with them. You can look back at the chapter text to refresh your memory or check the details of a reaction. Write your answers down and have a friend or parent check your score with the answers in the back of the book. You may want to linger in a chapter until you have a good handle on the material and get most of the answers right before moving on.

This book is divided into major sections. A multiple-choice test follows each of these sections. When you have completed a section, go ahead and take the section test. Take the tests "closed book" when you are confident about your skills on the individual quizzes. Try not to look back at the text material when you are taking them. The questions are no more difficult than the quizzes, but serve as a more complete review. I have thrown in several "wacky" answers to keep you awake and make the tests more fun. A good score is three-quarters of the answers right. Remember, all answers are in the back of the book.

The final exam at the end of the course is made up of easier questions than those of the quizzes and section tests. Take the exam when you have finished all the chapter quizzes and section tests and feel comfortable with the material as a whole. A good score on the exam is at least 75 percent of correct answers.

With all the quizzes, section tests, and the final exam you may want to have your friend or parent give you your score without telling you which questions you missed. Then you will not be as likely to memorize the answers to the questions you missed, but go back and see if you missed the point of the idea. When your scores are where you'd like them to be, go back and check the individual questions to confirm your strengths and areas that need more study.

Try going through one chapter a week. An hour a day or so will allow you to take in the information slowly. Don't rush. Chemistry is not difficult, but does take some thought. Just slug through at a steady rate. If you are especially interested in metals, spend more time on Chapter 12. If you want to learn the latest about nanotechnology, allow more time on Chapter 18. At a steady pace, you will complete the course in a few months. After completing the course and you have become a chemist-in-training, this book can serve as a ready reference guide with its comprehensive index, Periodic Table, and many examples of reactions and molecular bonding.

Suggestions for future editions are welcome.

Linda D. Williams

DEDICATION

This book is dedicated to the crew of the Space Shuttle *Columbia* (STS-107), Commander Rick Husband, Pilot William McCool, and Mission Specialists Dr. David Brown, Dr. Lauren Clark, Dr. Kalpana Chawla, Michael Anderson and Ilan Ramon for their strength, courage, and great sacrifice in advancing scientific knowledge for us all. Thank you.

Linda D. Williams

ACKNOWLEDGMENTS

The illustrations in this book were generated with Corel DRAW and Microsoft Powerpoint and Microsoft Visio courtesy of the Corel and Microsoft Corporations respectively.

I wish to express my thanks to Mary Kaser for her help with the technical editing of the manuscript for this book.

A very special thanks to Stan Gibilisco for timely encouragement.

My heartfelt thanks to my family and friends for their patience and faith.

Linda D. Williams

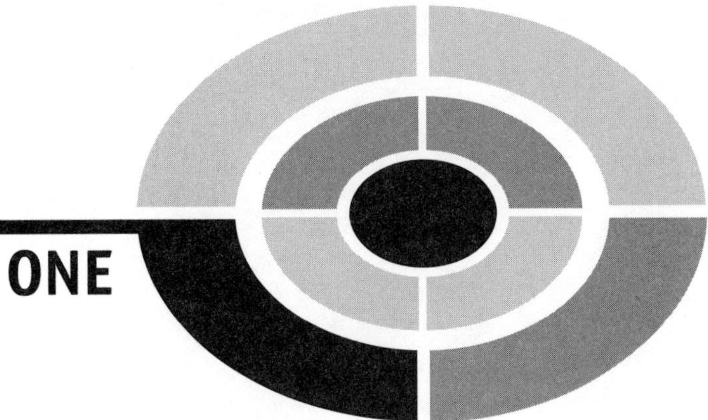

PART ONE

Understanding Matter

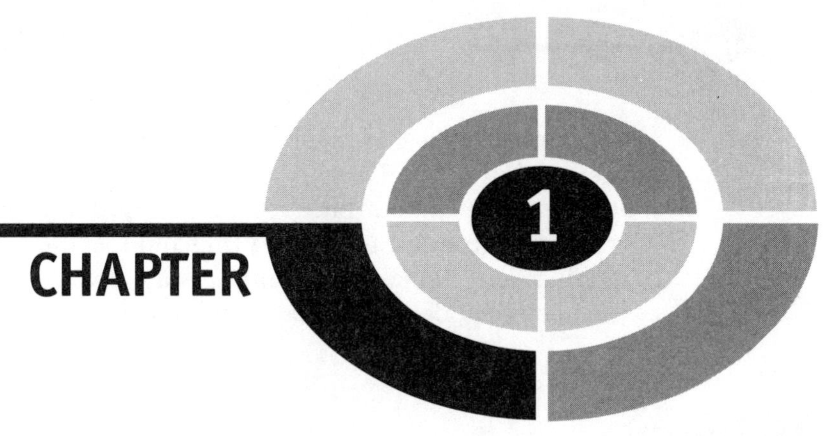

Scientific Method and Chemistry

Long ago, the first humans stood upright, used tools, and discovered that lightning produced fire. They found that the differences between medicinal extracts and plant toxins were slim and sometimes had very different effects. Everyday life included the drying of fish and meat with salt, the concentration of liquids into dyes, and the melting of metal ores to make tools. Scientific testing was sketchy. Trial and error provided clues to how elements, compounds, atoms, gases, and the like made up the world. What worked was carried over to the next generation; what didn't was discarded.

Aristotle (384–322 BC), a student at the Greek Academy, believed that matter was composed of four elements, fire, water, air, and earth. He did not think they were pure substances, but the solid, liquid, and gaseous forms of *proto hyle*, or *primary matter*. Aristotle wrote that neither matter nor form existed alone, but in combinations of hot, moist, dry, and cold, which united to form the elements. This explanation of the world was accepted for nearly 1800 years.

The main source of learning for much of the Western world until that time came from the Greeks and Romans. The strong desire to find out how the world worked kept classic philosophers pondering the mysteries of nature. However, during the Dark Ages (AD 500–1600), the growing knowledge of the time slowed quite a bit. Nomadic groups and barbarians from the cold north swept through Europe and England seeking conquest. People got busy protecting their homes and trying to stay alive.

Chemistry suddenly became very important.

Alchemy

Aristotle's four-element theory along with the formation of metals became the basis of early chemistry or *alchemy* as it was known then. A mixture of trickery and art, alchemy promised amazing things to those who held its power.

Alchemists were divided into two groups, *adepts* and *puffers*. Adepts considered themselves the true alchemists who could only produce gold through spiritual perfection. They called their attempts the Magnum Opus or Great Work.

In order to gain more prestige in the eyes of the rulers, adepts were initiated in stages. They had to move from one holy place to another seeking new methods and becoming enlightened. In order to move up the ladder of acclaim, they had to travel to the Chartres Cathedral in France or the Cathedral of St. James of Compostela in Spain. There they could feel the vibrations of the earth and experience spiritual transformation allowing them to achieve perfection and the power to create gold.

Puffers, on the other hand, pursued riches through the technique of *transmutation* and leaned toward showy, seemingly magical methods. Puffers got their name through the constant use of bellows in their practices. They used different types of furnaces and the ever-present bellows along with special fuels of oil, wax, pitch, peat, and animal dung. The common thought was that the hotter the fire, the quicker the transmutation.

The alchemists' two different paths led to widely different kinds of testing, but in the end, lots of ideas came from their efforts.

Melanosis, Leukosis, and Xanthosis

Alchemists thought color was a basic property of a metal. In their attempts to make gold, they decided that they had to first get the golden color. To do this, they performed a three-step color changing method called *melanosis*, *leukosis*, and *xanthosis*. These three steps in the method worked to bring a yellow-gold color to base metals of very different colors.

Melanosis, leukosis, and xanthosis alchemy procedures were performed in a device called a *kerotakis*. **Figure 1.1** shows how a kerotakis might have looked. In the kerotakis, a metal sample like copper was placed on a screen in a tall container. Sulfur was added and the entire container heated by a small fire in the base. When the sulfur was hot, it acted on the metal. Condensing sulfur-containing sulfides settled out. A sieve, like a spaghetti strainer only flat, held back pieces of unreacted metal, while a black compound collected at the kerotakis bottom. This mixture was heated in an open container to remove any extra sulfur.

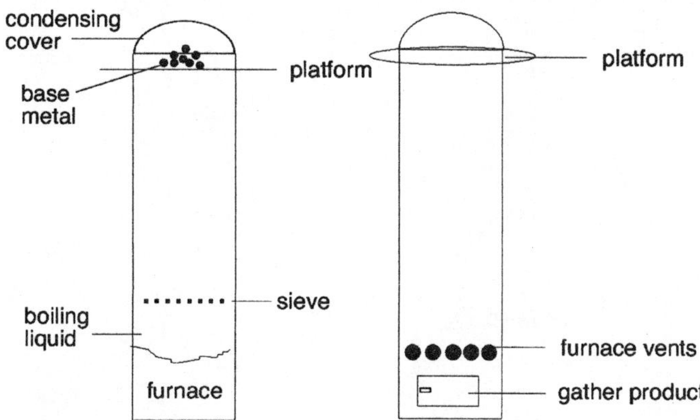

Fig. 1.1. A kerotakis device used a divided chamber with a furnace to provide heat.

The metal blackening was the first step in the process of melanosis. It was thought that a metal's original color was removed through this process. Since, melanosis darkened the metal's original color, alchemists thought they had banished it.

Following the blackening stage, another compound like arsenic sulfide was added to whiten the copper. This second step was called leukosis.

The third and final color-changing step, xanthosis, called for the addition of a calcium polysulfide solution (usually made of lime, sulfur, and vinegar). Mixed with the whitened copper, the solution was heated until the metal's surface took on a tinted, yellow color. This step gave the golden color that alchemists bragged was the newly formed gold.

Alchemists thought that heating base metals with sulfur caused the freeing of gold from a metal. They thought that when they got the golden color, they had gold. Since a lot of the rulers didn't know any better, the newly made gold gained alchemists acceptance for a while.

Other alchemists, eager to please those in power, thought they could create gold from other base metals such as lead and zinc. Ambitious rulers, looking for ways to fund their war machines, sponsored many of these early attempts.

Alchemists became the new superstars. Those who made wild claims they couldn't deliver, were permanently benched. Others made progress. Crystallization and distillation of solutions began to be understood and used as standard practice. Many previously unknown elements and compounds were discovered.

Alchemists often used the image of a serpent catching its own tail as a way to symbolize the unity and convertibility of the elements. Early alchemists used the signs of the planets, to which they thought the elements were connected, as symbols for metals. This is illustrated in **Figure 1.2.**

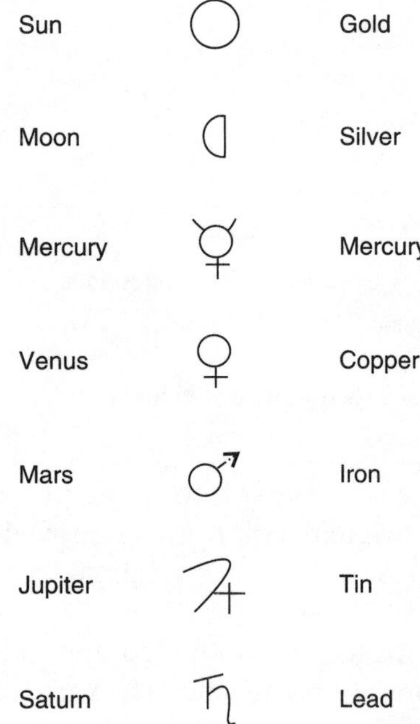

Fig. 1.2. Symbols of the planets were used to identify metals.

Other areas of science were advancing now, and in 1543 Nicolas Copernicus made a *hypothesis* based on his observations of the planets. He thought that the Earth and planets rotated through space around the Sun, not the Earth, as was commonly believed at that time.

A **hypothesis** is a statement or idea that describes or attempts to explain observable information.

Copernicus believed that from the Sun outwards rotated Mercury, Venus, Earth (with the moon rotating around it), Mars, Jupiter, and Saturn. This strange, new hypothesis wasn't well accepted since everyone knew that the Sun revolved around the Earth. Even the alchemists wondered how different metals might be affected.

An **experiment** is a controlled testing of the properties of a substance or system through carefully recorded measurements.

In 1609, Galileo Galilei tested Copernicus' hypothesis with a home-built telescope (there were no factories then). He took measurements and recorded data that confirmed Copernicus' hypothesis. Galileo discovered the key to valid research, *experimentation*. Curious about how things worked, he recorded his observations with respect to changing factors such as time, angle in the sky, and position of the Moon, Sun, and stars. His observations and calculations led to the discovery of the four satellites of Jupiter in 1610. As a result of his experiments, Galileo is thought of as the founder of the *scientific method*.

Antoine Lavoisier (1743–94) insisted on accurate measurements (which we will discuss more in Chapter 2) and developed a theory of combustion. He determined that combustion results from a chemical bonding between a burning substance and a component of the air (which he named oxygen), to form something new.

A **theory** is the result of thorough testing and confirmation of a hypothesis. A theory predicts the outcome of new testing based on past experimental data.

Lavoisier found that liquid mercury when burned in the air became a red-orange substance with a greater mass than that of the original mercury. He also showed that the original mass of mercury could be regained when the new substance was heated.

$$\text{mercury} + \text{oxygen} \Rightarrow \text{mercuric oxide}$$

Along with experiments by Joseph Priestly, Lavoisier discovered that the air was composed of several different components, including nitrogen, instead of one all-purpose gas. Curious about what was in the air that added to combustion, he performed experiments with other gases. These experiments showed that nitrogen did not support combustion even though it was a component of "air."

In experiments with water, Lavoisier found that water contains hydrogen and oxygen. He was also the first person to arrange chemicals into family groups and to try to explain why some chemicals form new compounds when mixed. Due to his experiments, Lavoisier is said to be the father of *modern chemistry*.

Following experimentation in many fields such as astronomy, electricity, mathematics, biology, chemistry, and medicine, data were recorded that showed how nearly everything could be studied and predicted through a series of successive observations and calculations. When the same results were repeatedly obtained by a variety of experimenters in different laboratories in various countries, a particular hypothesis or theory became a *law*.

> A **law** is a hypothesis or theory that is tested time after time with the same resulting data and thought to be without exception.

John Dalton developed the *law of partial pressures* in 1803. Dalton, interested in the Earth's atmosphere, recorded more than 200,000 atmospheric findings in his notebooks. These observations prompted Dalton to study gases and from the results of his experiments explained the condensation of dew and developed a table of the vapor pressures of water at different temperatures.

By extending these experiments, Dalton proved that the total pressure of a gas in a system is equal to the sum of the partial pressures of each constituent gas ($P_{total} = P_1 + P_2 + P_3 + \ldots$). He was also the first to publish the generalization that all gases, initially at the same temperature, expand equally as they increase in temperature.

Atomic Theory

In 1803, Dalton began to formulate his most important contribution to science, the *atomic theory*. While examining the nitrogen oxides and the percentage of nitrogen found in the air, he noted the interaction of nitric oxide with oxygen. He found that the reaction seemed to occur in two different proportions with the same exact ratios:

$$2NO + O \rightarrow N_2O_3$$

$$NO + O \rightarrow NO_2$$

Dalton noticed that oxygen combined with nitrogen in a ratio of 1 to 1.7 and 1 to 3.4 by weight. After testing this observation many times, he proposed the *law of multiple proportions*, where element weights always combine in small whole number ratios. Dalton published his initial list of atomic weights and symbols in the summer of 1803, which formally gave chemistry the vocabulary (symbol names) that we have come to know and memorize.

Moreover, Dalton's most famous work, *A New System of Chemical Philosophy, Part I*, enlarged the idea that no two compound fluids have the same number of particles or the same weight. Dalton relied on his experimental and mathematical hypotheses to cobble together a previously unthinkable theory. He reasoned that atoms must combine in the simplest possible configurations in order to be consistently the same. It seemed straightforward then, to use the idea of individual atoms and particles when showing various chemical reactions.

The law of partial pressures, along with laws proposed by such scientists as Robert Boyle, Jacques Charles, and Joseph Gay-Lussac increased the growing body of scientific knowledge that believed that all components of nature such as gases, pressure, and heat were interconnected. We will discuss these laws in detail in Chapter 17.

Applied Science

Matter is the basic material of which things are made. Chemists discover new elements and further define the amazing properties of matter every day. They keep finding creative uses for compounds unknown thirty or forty years ago.

The *National Aeronautical and Space Administration* (NASA), for example, is famous for applying basic science in new ways.

NASA uses the scientific method to perform applied science. They see how something behaves in space with almost no gravity, like the formation of crystals, and then look for ways that the same application can be used in ground-based experiments. By teaming with scientists in industry, NASA improves pharmaceuticals, optics, and bioengineering devices. Research applied in this way can more quickly travel from the laboratory to the individual.

At NASA, these dual-purpose science and technology brainstorms are called *spinoffs*. A sampling of NASA's Science and Technology Spinoffs is provided in **Table 1.1**. NASA spinoffs include computer technology, consumer/home/recreation products, environmental and resource management, industry and manufacturing, public safety, and transportation.

Table 1.1 NASA spinoffs are applications of basic science.

- Bioreactor—a cell culture device developed at NASA-Johnson Space Center that brings a new scientific tool to cancer and virus testing without risking harm to patients. The rotating bioreactor wall allows three-dimensional growth of tissues without limiting pressure points. It has been successful in culturing over 35 cell types.
- Ultrasound Skin Damage Assessment—enables immediate assessment of burn damage depth and course of treatment.
- Low Vision Enhancement System (LVES)—provides a video scene via a system of optical mirrors that project video images onto the wearer's retinas. The headset, worn like aviators' goggles, helps counteract the effects of macular degeneration associated with aging, diabetic retinopathy, glaucoma, and tunnel vision.
- Heart Rate Monitor—through the use of a thin dielectric film, this dry reusable electrode allows contact that is not affected by heat, cold, light, perspiration, or rough or oily skin. It permits precise heart rate monitoring for cardiac rehabilitation patients as well as professional athletes.
- Medical Gas Analyzer—astronaut physiological monitoring technology. When used to measure operating room anesthetic concentrations such as oxygen, carbon dioxide, and nitrogen, it ensures precise breathing environments for surgery patients.

The keys to the scientific method are curiosity and determination, observation and analysis, measurement, and conclusion. As humans, we are curious by nature. In the following chapters, you will learn how scientists satisfy their curiosity.

Quiz 1

1. What was the first major demonstration of a chemical reaction that produced heat?
   (a) mold
   (b) fire
   (c) ice
   (d) earthquake

2. During the Dark Ages, alchemists
   (a) promised to turn lead into gold
   (b) were the first true experimenting chemists
   (c) discovered crystallization and distillation procedures
   (d) all of the above

3. A hypothesis is a
   (a) container for performing experiments
   (b) way to describe heat transfer between minerals
   (c) sterile medical device
   (d) statement or idea that describes or attempts to explain observable information

4. Which early scientist accurately described the configuration of the Sun, Moon, and planets in relationship to each other?
   (a) Linus Pauling
   (b) Claudius Ptolemy
   (c) Nicolas Copernicus
   (d) Leonardo da Vinci

5. An experiment is
   (a) a controlled testing of the properties of a substance or system through carefully recorded measurements
   (b) an uncontrolled testing of the properties of a substance or system through recorded measurements
   (c) a one-time reporting of a few observable characteristics
   (d) a bad choice brought on by peer pressure

6. Who is said to be the founder of the scientific method?
   (a) Alexander Fleming
   (b) Joseph Priestly
   (c) Galileo Galilei
   (d) Antoine Lavoisier

7. A theory
   (a) accounted for a ruler's need to produce gold from zinc
   (b) is the result of sudden inspiration during a lightning storm
   (c) predicts the outcome of new testing based on past experimental data
   (d) is a type of atomic particle

8. John Dalton proposed the first theory on
   (a) the rotation of the satellites around Saturn
   (b) the characteristics of individual atoms and particles

(c) the complex interactions of solids when melted
(d) the neutralization of pH

9. A scientific law is best described as
(a) a series of rules made by representatives of the government
(b) a good idea that many people agree with voluntarily
(c) the transmutation of lead into gold
(d) a hypothesis or theory that is tested repeatedly with the same results and thought to be without exception

10. The law of partial pressures can be best described by the following equation:
(a) $P_{total} = P_1 + P_2 + P_3$
(b) $P_{total} = (P_1 + P_2)/P_3$
(c) $P_{total} = (P_1 + P_2)\mu$
(d) $P_{total} = 2(P_1 \times P_2 \times P_3)$

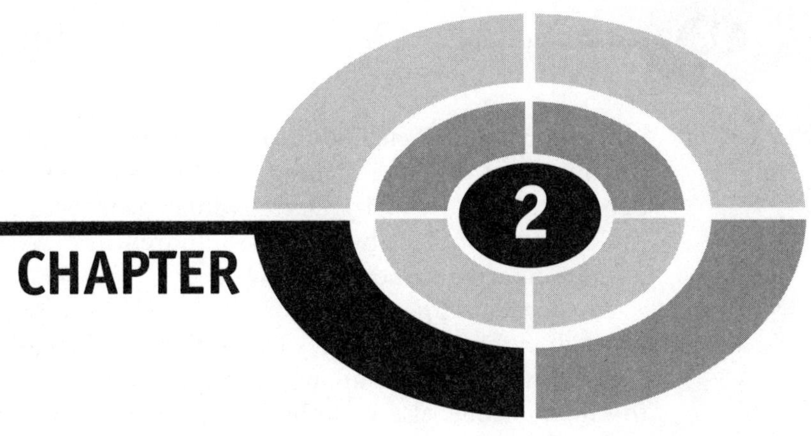

Data and How to Study It

Experimental Science

Chemistry is an *experimental* science. It is divided into two branches, *pure chemistry* and *applied chemistry*. Pure chemistry is theoretical and predicts results of experiments or observations. Applied chemistry involves the practical applications of materials and reactions. How is rust formed and how do you remove it? How can clothes get clean from washing them with soap made from ashes and fat? Why does copper turn green and then black when exposed to air? How can computer chips made from sand (silicon) carry information and electricity?

Measurements

Observation and measurement, as in all science, are the keys to chemistry. In research, as in other parts of life, we are constantly measuring using common units. The baseball cleared the outfield fence by a foot. The soccer ball missed the flowerpot by three inches. The Austrian driver cruised at 160 kilometers per hour. The Kentucky Derby winner won by a length. The Olympic skier pulled into first place by two one-hundredths of a second. The soldier's letter home weighed 1 ounce.

A *chemical experiment* is a controlled testing of a sample's properties through carefully recorded observations and measurements.

Research is all about measuring. However, to repeat an experiment or follow someone else's method, the **same** *units* must be used. It wouldn't work to have a researcher in New York measuring in cups while another in Germany measured in milliliters. To repeat an experiment and learn from it, scientists around the world needed a common system.

In 1670, a French scientist named Gabriel Mouton suggested a *decimal system* of measurement. This meant that units would be based on groups of ten. It took a while for people to try it for themselves, but in 1799 the French Academy of Sciences developed a decimal-based system of measurement. They called it the *metric system*, from the Greek word *metron*, which means a measure. On January 1, 1840, the French legislature passed a law requiring the metric system be used in all trade.

International System of Units (SI)

In 1960, the General Conference on Weights and Measures adopted the *International System of Units* (or SI, after the French, Le Systeme International d'Unites). The International Bureau of Weights and Standards in Sievres, France, houses the official platinum standard measures by which all other standards are compared. The SI system has *seven base units* from which other units are calculated. **Table 2.1** gives the SI units used in chemistry.

When Great Britain formally adopted the metric system in 1965, the United States became the only major nation that didn't require metric, though people had been using it since the mid-1800s.

The advantage of the SI system is that it is a measuring system based on a decimal system. With calculations written in groups of ten, results can be easily recorded as something called *scientific notation*. There are written prefixes that indicate exponential values as well. Some of these are listed in **Table 2.2** which lists terms used in scientific notation.

Table 2.1 SI base units are used in chemistry.

Measurement	**Unit**	**Symbol**
Mass (not weight)	kilogram	kg
Length	meter	m
Temperature	kelvin	K
Time	second	s
Pure substance amount	mole	mol
Electric current	ampere	A
Light brightness (wavelength)	candela	cd

Exponential or Scientific notation is a way of writing numbers as powers of ten.

Scientific Notation

Scientific notation is a simple way to write and keep track of large and small numbers without a lot of zeros. It provides a short cut to recording results and doing calculations. The ease of this method is shown below.

EXAMPLE 2.1

$$100 = (10)(10) = 10^2 = \text{one hundred}$$

$$1{,}000 = (10)(10)(10) = 10^3 = \text{one thousand}$$

$$10{,}000 = (10)(10)(10)(10) = 10^4 = \text{ten thousand}$$

$$100{,}000 = (10)(10)(10)(10)(10) = 10^5 = \text{one hundred thousand}$$

$$1{,}000{,}000 = (10)(10)(10)(10)(10)(10) = 10^6 = \text{one million}$$

$$1{,}000{,}000{,}000 = (10)(10)(10)(10)(10)(10)(10)(10)(10) = 10^9 = \text{one billion}$$

Table 2.2 Scientific notation helps determine the scale of measurements.

Prefix	Symbol	Value
tera	T	10^{12}
giga	G	10^{9}
mega	M	10^{6}
kilo	k	10^{3}
deca	da	10^{1}
deci	d	10^{-1}
centi	c	10^{-2}
milli	m	10^{-3}
micro	μ*	10^{-6}
nano	n	10^{-9}
pico	p	10^{-12}

*Greek letter mu

$$1/10 = 10^{-1} = \text{one tenth}$$

$$1/100 = 1/(10)(10) = 10^{-2} = \text{one hundredth}$$

$$1/1{,}000 = 1/(10)(10)(10) = 10^{-3} = \text{one thousandth}$$

$$1/10{,}000 = 1/(10)(10)(10)(10) = 10^{-4} = \text{one ten thousandth}$$

$$1/1{,}000{,}000 = 1/(10)(10)(10)(10)(10)(10) = 10^{-6} = \text{one millionth}$$

$$1/1{,}000{,}000{,}000 = 1/(10)(10)(10)(10)(10)(10)(10)(10)(10) = 10^{-9} = \text{one billionth}$$

Since the *English System* was used in the United States for many years with units of inches, feet, yards, miles, cups, quarts, gallons, etc., many people were not comfortable with the metric system until recently. Most students wonder why they ever preferred the older system, when they discover how easy it is to multiply metric units.

Table 2.3 lists some everyday metric measurements.

Table 2.3 Metric measurements can describe different scale objects.

Sample	Measurement (meters)
Diameter of uranium nucleus	10^{-13}
Water molecule	10^{-10}
Protozoa	10^{-5}
Earthworm	10^{-2}
Human	2
Mount Everest	10^{3}
Diameter of the Earth	10^{7}
Distance from Pluto to the Sun	10^{13}

Significant Figures

Measurements are never exact, but scientists try to record an answer with the least amount of uncertainty. This is why the idea of scientific notation was set up, to standardize measurements with the least uncertainty. The idea of *significant figures* was used in order to write numbers either in whole units or to the highest level of confidence.

Significant figures are the number of digits written after the decimal point to measure a quantity.

A *counted* significant figure is something that cannot be divided into subparts. These are recorded in whole numbers such as 9 chickens, 2 bicycles, or 7 keys. *Defined* significant digits are exact numbers, but not always whole numbers, like 2.54 centimeters equals one inch.

EXAMPLE 2.2
How many significant figures are in the following?

(a) 9.107 (4, zero in the middle is significant)
(b) 401 (3, zero in the middle is significant)
(c) 0.006 (1, leading zeros are never significant)
(d) 800 km (3, zeros are significant in measurements unless otherwise indicated)
(e) 3.002 m (4, zeros in the middle of non-zero digits are significant)

When finding the number of significant figures, the easiest shortcut is to look at the zeros acting as placeholders.

Leading zeros at the beginning of the left-hand side of a number are never significant. You start at the left and count to the right of the decimal point. The measurement 0.096 m has two significant figures. The measurement 13.42 cm has four significant figures. The mass 0.0027 g has two significant figures. (Note: remember to leave off the leading zeros.)

Sandwiched (in the middle) zeros are always significant. The number 26,304 has five significant figures. The measurement 0.000001002 m has four significant figures.

Scientific notation gets rid of guessing and helps to keep track of zeros in very large and very small numbers. If the diameter of the Earth is 10,000,000 m, it is more practical to write 1×10^7 m. Or, if the length of a virus is 0.00000004 m, it is easier to write 4×10^{-8} m.

When multiplying or dividing numbers, the significant digits of the number with the least number of significant digits gives the number of significant digits the answer will have.

EXAMPLE 2.3
40 lb potatoes × \$0.45 per lb = \$18.00 or \$18 since the first number is only measured to two places.

EXAMPLE 2.4
0.5 ounce of perfume × \$25.00 per ounce = \$12.50 for 0.5 ounces of perfume. (Note: the zero is only written because you cannot divide coins further.)

EXAMPLE 2.5
6.23 ft. of wood × \$2.00 per linear foot = \$12.50 per linear foot.

Imagine you are trying to prove a theory based on a specific property, like boiling point. Unless the boiling point temperature was recorded with precision by other chemists, you will have trouble repeating the experiment, let alone proving a new theory. Since theories become laws by repeated experimentation, it is important to record measurements precisely.

Scientific knowledge moves forward by building upon results and experiments done by earlier scientists. If measurements are taken with little care or precision, a researcher doesn't know if the observed results are new and exciting or just plain wrong.

Precision is the closeness of two sets of measured groups of values.

Precision is directly related to the amount of reproducibility of a measurement. Closely related to the topic of precision is that of *accuracy*. Some people use the two interchangeably, but there is a difference.

Accuracy is linked to how close a single measurement is to its true value.

In baseball, when Player A throws balls at a target's center, it represents high precision and accuracy. Player B's aim, with balls high and low missing the target, represents low precision and low accuracy. Player C's hits, clumped together at the bottom left side of the target, define high precision (since they all landed in the same place), but low accuracy (since the object is to hit the target's center). Player C, then, has to work on hitting the target's center, if he wants to win games and improve accuracy.

ROUNDING

Rounding is the way to drop (or leave off) non-significant numbers in a calculation and adjusting the last number up or down. There are three basic rules to remember when rounding numbers:

(1) If a digit is ≥ 5 followed by non-zeros, then add 1 to the last digit. (Note: 3.2151 would be rounded to 3.22.)
(2) If a digit is < 5 then the digits would be dropped. (Note: 7.12132 would be rounded to 7.12.)

(3) If the number is 5 (or 5 and a bunch of zeros), round to the least certain number of digits. (Note: 4.825, 4.82500, and 4.81500 all round to 4.82.)

Rounding reduces accuracy, but increases precision. The numbers get closer, but are not necessarily on target.

EXAMPLE 2.6
Try rounding the numbers below for practice.

(a) 2.2751 to 3 significant digits
(b) 4.114 to 3 significant digits
(c) 3.177 to 2 significant digits
(d) 5.99 to 1 significant digit
(e) 2.213 to 2 significant digits
(f) 0.0639 to 2 significant digits

Did you get (a) 2.28, (b) 4.11, (c) 3.2, (d) 6, (e) 2.2, and (f) 0.064?

When multiplying or dividing measurements, the number of significant digits of the measurement with the least number of significant digits determines the number of significant digits of the answer.

EXAMPLE 2.7
Do you see how significant digits are figured out?

(1) 1.8 pounds of oranges × \$3.99 per pound = \$7.182 = \$7.18 = \$7.2 (Note: 1.8 pounds of oranges has two significant digits)
(2) 15.2 ounces of olive oil × \$1.35 per ounce = \$20.50
(3) 25 linear feet of rope × \$3.60 per linear feet = \$90.00

Measurements can be calculated to a high precision. Calculators give between 8 and 10 numbers in response to the numbers entered for a calculation, but most measurements require far less accuracy.

Rounding makes numbers easier to work with and remember.

Think of how out-of-town friends would react if you said to drive 3.4793561 miles west on Main Street; turn right and go 14.1379257 miles straight until Union Street; turn left and travel 1.24900023 miles around the curve until the red brick house on the right. They might never arrive! But by rounding to 3.5 miles, 14 miles, 1.2 miles, and watching the car's odometer (the instrument that measures distance), they would arrive with a lot less trouble and confusion.

Conversion Factors

Conversion factors make use of the relationship between two units or quantities expressed in fractional form. The *factor-label method* (also known as *dimensional analysis*) changes one unit to another by using conversion factors.

Conversion factors are helpful when you want to compare two measurements that aren't in the same units. If given a measurement in meters and the map reads only in kilometers, you have a problem. You could guess or use the conversion factor of 1 km/10^3 m. Look at the conversion below.

$$0.392 \text{ m} \times 1 \text{ km}/10 \text{ m}^3 = 0.392 \times 10^3 \text{ km}$$

$$= 3.92 \times 10^{-4} \text{ km}$$

If you have centimeters and need to know the answer in inches, then use the conversion factor 1 inch/2.54 cm.

914 cm × 1 inch/2.54 cm = 360 inches (since 914 has 3 significant digits)

Converting measurements can also be a two-step process.

$$\text{mg} \Rightarrow \text{g} \Rightarrow \text{kg}$$

$$\text{liters} \Rightarrow \text{quarts} \Rightarrow \text{gallons}$$

$$\text{miles per hour} \Rightarrow \text{meters per minute}$$

Look at the two-step conversions below.

EXAMPLE 2.8

$$2461 \text{ mg} \Rightarrow \text{kg}$$

$$\text{mg} \Rightarrow \text{g} \Rightarrow \text{kg}$$

$$1 \text{ mg} = 10^{-3} \text{ g}; 1 \text{ kg} = 10^3 \text{ g (conversion factors)}$$

$$2461 \text{ mg} \times 10^{-3} \text{ g/mg} \times 1 \text{ kg}/ 10^3 \text{ g}$$

$$= 2461 \times 10^{-6} \text{ kg} = 2.461 \times 10^{-3} \text{ kg}$$

EXAMPLE 2.9

$$8.47 \text{ liters} \Rightarrow \text{gallons}$$

$$\text{liters} \Rightarrow \text{quarts} \Rightarrow \text{gallons}$$

$$1.06 \text{ qt/liters; } 1 \text{ gal}/4 \text{ qt (conversion factors)}$$

$$8.47 \text{ liters} \times 1.06 \text{ qt/liter} \times 1 \text{ gal}/4 \text{ qt} = 2.24 \text{ gallons}$$

EXAMPLE 2.10

$$70 \text{ miles/hour} \Rightarrow \text{meters/minute}$$

$$\text{miles/hr} \Rightarrow \text{km/hr} \Rightarrow \text{m/hr} \Rightarrow \text{m/min}.$$

$$1.61 \text{ km/mi; } 10^3 \text{ m/km; } 1 \text{ hr/60 min (conversion factors)}$$

$$70 \text{ mi/hr} \times 1.61 \text{ km/mi} \times 10^3 \text{ m/km} \times 1 \text{ hr/60 min} = 1878.33 \text{ m/min} = 1.9 \times 10^3 \text{ m/min}$$

SI derived units are obtained by combining SI base units.

Temperature

The measure of the intensity of heat of a substance is said to be its *temperature*. A thermometer measures temperature. Temperature is measured in three different units: *Fahrenheit* (°F) in the United States, *Celsius* (°C) in science and elsewhere, and *Kelvin* to measure absolute temperature. **Figure 2.1** shows how the temperature systems compare.

Fahrenheit	Celsius	Kelvin
212	100	373
176	80	353
140	60	333
104	40	313
68	20	293
32	0	273
−4	−20	253
−40	−40	233

Fig. 2.1. A comparison of the three temperature scales shows their differences clearly.

The conversion factor for Celsius to Fahrenheit is

$$t(\mathrm{F}) = [t(\mathrm{C}) \times 1.8\mathrm{F}/1^\circ\mathrm{C}] + 32 = [t(\mathrm{C}) \times 1.8] + 32$$

The conversion factor for Fahrenheit to Celsius is (*hint*: subtract 32 so that both numbers start at the same temperature)

$$t(C) = [t(F) - 32°F] \times 1°C/1.8°F = [t(F) - 32]/1.8$$

or a simpler way to state it is:

$$°C = 5/9\ (°F - 32)$$

EXAMPLE 2.11

A summer day in Hawaii might be 21°C. What is that in Fahrenheit?

$$21°C = 5/9\ (°F - 32)$$

$$21 + 32 = 5/9°F$$

$$53 \times 9 = 5°F$$

$$477/5 = 70°F$$

To obtain *absolute zero* (the lowest temperature possible), the *kelvin* scale is used, where the lowest temperature is zero. A kelvin is a SI temperature unit. The heat energy is zero.

To see how temperature conversion works, let's convert normal body temperature, 98.6°F, to Celsius.

EXAMPLE 2.12

$$°C = 5/9\ (°F - 32)$$

$$°C = 5/9\ (98.6°F - 32)$$

$$= 5/9\ (66.6) = 37.0°C$$

°C can be converted to K by adding 273 to the Celsius temperature.

EXAMPLE 2.13

$$K = °C + 273$$

$$K = 37°C + 273 = 310\ K$$

In later chapters, we will study reactions where heat can play an important role in determining the character of the final compound.

Quiz 2

1. Chemistry is known as
   (a) an attraction between two people
   (b) an exact science
   (c) an experimental science
   (d) a method to describe units of heat

2. In 1670, Gabriel Mouton suggested
   (a) a law of partial pressures
   (b) the boiling point of alcohol
   (c) the Sun as the center of the universe
   (d) a decimal system of measurement

3. The International System of Units (SI) has how many base units?
   (a) 4
   (b) 6
   (c) 7
   (d) 9

4. Exponential or scientific notation is
   (a) a method where numbers are written in powers of 10
   (b) a shorthand method of number accounting
   (c) a way to write very large and very small numbers
   (d) all of the above

5. The number of digits recorded in a measurement is
   (a) always whole numbers
   (b) significant digits or figures
   (c) a way to count on your fingers
   (d) the method of including all zeros

6. Precision is described as
   (a) more accurate than excision
   (b) less accurate than two significant digits
   (c) the closeness of two sets of measured groups of values
   (d) the equal spacing of numbers around a common number

7. Accuracy is described as
   (a) more precise than two significant digits
   (b) the closeness of two sets of measured groups of values
   (c) only applicable to experimental measurements
   (d) the closeness of a single measurement to its true value

8. Rounding is used primarily to
   (a) sum up significant figures
   (b) drop non-significant digits in a calculation
   (c) drop digits greater than 5
   (d) increase all numbers to the most certain number

9. Conversion factors make use of
   (a) a relationship between two units or quantities in fractional form
   (b) the fact that units are always written as whole numbers
   (c) numbers which cannot be divided into smaller units
   (d) a direct connection between weight and volume

10. Which of the units below is an example of SI derived units?
   (a) cm/m
   (b) m/s^2
   (c) m/kg^2
   (d) m/ft^2

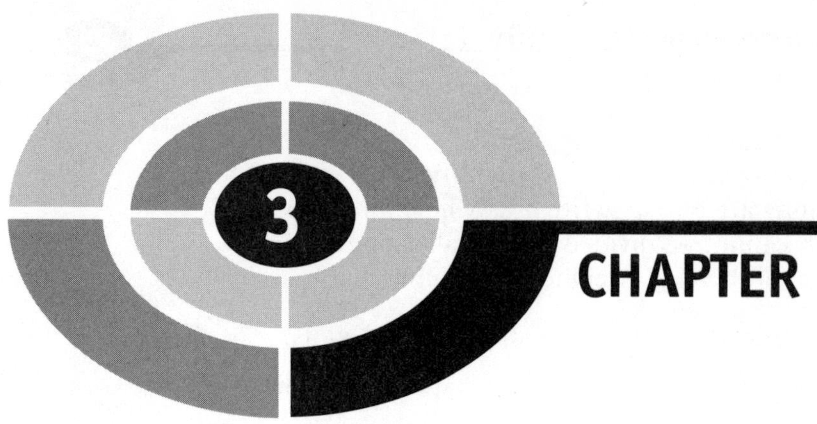

Properties of Matter

Matter

Chemistry is defined as the study of matter and the way it reacts in different situations. But, what is matter?

At first, people thought matter only included things that could be seen and measured, like salt, grain, and olive oil. Everyone used their senses to define what they saw, heard, tasted, touched, and smelled. Anything that could not be sensed simply did not exist or belonged in the realm of myths and legends. Most people only believed in what they saw for themselves. They doubted everything until they had experienced it for themselves, usually by seeing it "with their own eyes."

With the beginning of the scientific era, the search for exactly what makes up the "stuff" of the universe became more intense.

> **Matter** is defined as anything that has mass and occupies space.

We now know that even in the unseen world matter exists that is too small to be seen or measured except with very complex machines and sometimes not even then. Sometimes all scientists can do is observe the effects of matter, even though the actual matter cannot be obtained.

Atomic Theory

Around 495 BC, a Greek philosopher named Democritus wondered if substances could be divided into smaller and smaller parts indefinitely. He thought that eventually particles would be reached that could no longer be divided. He called these smallest particles *atoms* (from the Greek word *atomos*, which means "not divided"). The great philosophers Aristotle and Plato thought matter was continuous, fluid, and could not be divided into individual particles.

In 100 BC, another forward thinker named Lucretius wrote a long, descriptive poem called *De Rerum Natura* (The Nature of Things), praising early ideas that leaned toward an *atomic theory* of matter. Since few people could read and most gained information through story telling, the poem helped people to understand the basic nature of matter. The invention of the Gutenberg printing press in 1452, which used olive oil ink with a screw-type wine press, helped to spread the knowledge of the time.

The De Rerum Natura poem was one of the first texts set in print. This fact helped the atomic theory survive. Along with religious texts and Bibles of the time, the poem was one of the few things available to read.

Once the first seeds of the atomic theory spread, scientists began thinking about matter in particle form. Experiments were performed and measurements taken to discover how compounds could be further divided, rearranged, and combined.

In the late 1700s, Antoine Lavoisier, the father of modern chemistry, insisted on precise measurements to better compare results and explain the properties of matter. Unfortunately, though a brilliant scientist, he was also associated with French taxation and the ruling governmental class. In 1794, his research was cut short by the guillotine and the French Revolution.

Since then, modern scientists have discovered particles that are smaller than an atom. These *subatomic* particles, which exist in the core of an atom, are called *protons* with a positive charge, and *neutrons* which are neutral and have no charge. *Electrons* orbit the nucleus like untamed satellites that are attracted by the forces of electromagnetism.

Figure 3.1 shows how the core of an atom might look if a model were made of the subatomic particles.

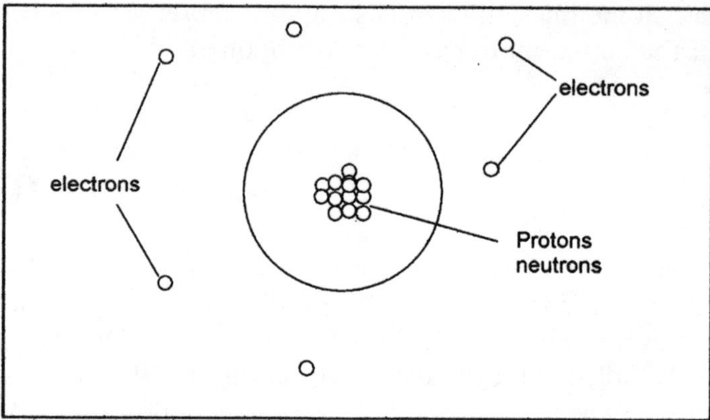

Fig. 3.1. The atomic nucleus is thought to have orbiting electrons.

In 1968, *quarks* were discovered. *Quarks* are particles of matter that are constituents of neutrons and protons. So far, six different types of quarks have been identified. These will be discussed more in later chapters.

Solids, Liquids, and Gases

Chemistry and the study of matter focus on the forms of matter. Scientists describe the three basic forms of matter as: *solid*, *liquid*, and *gas*. **Figure 3.2** illustrates these three different forms that matter can take.

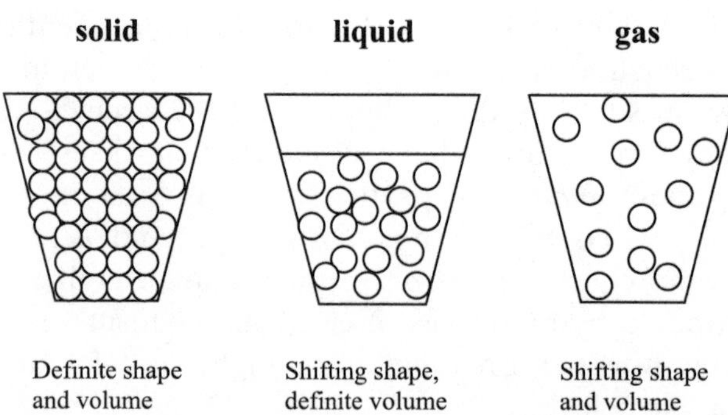

Fig. 3.2. The three forms of matter consist of solids, liquids, and gases.

Solids include things like boulders, metals, crystals, and glass. They are fixed in shape and rigid with a measurable volume. To change shape in a major way, a strong outside force like fire or heavy impact is needed.

Liquids such as water, oil, and alcohol have been known for centuries. They have a measurable volume, but are bendable and can change shape. Liquids are easier to study since they fill the shape of their container and flow from one place to another. They do not require force to change shape, but are affected by heat.

Gases are a different story. They have neither form nor volume and expand to fill the entire container into which they are placed. At times they are visible and then disappear. They seem to come from nothing and leave to go nowhere. Ancient alchemists' ideas of transmutation seem to better describe the mystery of gases.

To make things even more interesting, some liquids like water can take on all three modes (water, ice, and steam) at different times. Scientists wanted to understand how this happened. Through much experimentation, it was discovered that matter had specific *properties*.

Properties

The unique character or the way an element reacts is said to be its *properties*. These properties are grouped into two classes, *physical* and *chemical*.

Physical properties are special characteristics that make up the physical composition of a sample. Physical properties include: color, form, density, thermal and electrical conductivity, and melting and boiling points. Physical properties can be seen without any change in form or dimension. For example, barium is commonly silvery white in color, a solid, melts at 727°C (1341 °F) and burns green at a wavelength of 554 nanometers. Even melted, it is still barium, but in a different form than when it was solid.

Chemical properties are those characteristics that focus on a substance's behavior when mixed with another element or compound. For example, copper when exposed to oxygen turns green on the surface, but doesn't dissolve in water. The reaction that causes the surface to turn green forms a very thin layer (copper carbonate or copper sulfate) that actually protects the surface from further corrosion.

Iron is abundant in the Earth's crust and was one of the first refined metals. However, it is never found in its pure form, but as an oxide (combined with oxygen). Iron chemically reacts with air and water to form rust, a porous crumbling material that sticks loosely to the iron's surface. When rust crumbles and falls away, a new place on the iron's surface is exposed; eventually the entire sample will rust away.

Elements

Over time, scientists discovered that some matter is composed of *pure* chemicals. Pure substances are *homogeneous* and have certain unchanging chemical compositions. For example, a pure sample of highly condensed carbon, diamond, will always have the same crystalline structure. The repeating structural unit of diamond consists of eight atoms in cubic shapes. Using this cubic form and its highly symmetrical arrangement of atoms, diamond crystals form several different shapes. We will discuss this in more detail in Chapter 15. This cubic form and its light reflectivity make diamond one of the most desired substances on Earth.

Chemicals that have the same type of matter all through the sample are said to be pure *elements*. Oxygen, potassium, mercury, and nickel are pure elements.

During his research in 1789, Antoine Lavoisier defined an *element* as a substance that could not be decomposed by a chemical reaction into simpler substances. Lavoisier identified 33 elements that he thought were pure and indivisible. Of those 33, 20 of the 109 elements currently identified, are still considered pure elements.

> An **element** is made up of a pure sample with all of the same kinds of atoms and cannot be further separated into simpler elements.

Mixtures

Mixtures can be separated into two or more substances manually. No chemical reaction is needed. In nature, salt water can be separated into its components of water and salt by allowing the water to evaporate. Mixtures are found in two forms: *heterogeneous* and *homogeneous*.

A *heterogeneous* mixture is one with physically separate parts that have different properties. An easy example is salt and pepper. A heterogeneous mixture has separate *phases*. A *phase* represents the number of different homogeneous materials in a sample. Salt is all one phase and pepper is one phase. They do not have a wide variety of characteristics, but are physically separate.

A *homogeneous* solution has one phase (liquid) but may have more than one component within the sample. Again, salt water is an example of a homogeneous mixture. It is the same throughout, but has two parts: water and salt. **Figure 3.3** compares matter and its different parts.

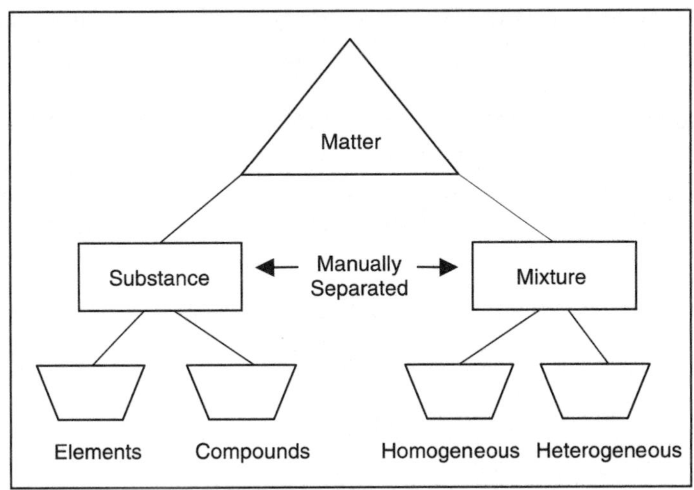

Fig. 3.3. Matter can be further broken down into different divisions.

Compounds

Pure chemicals that can be broken down into simpler chemicals are known as *compounds*. Commonly, chemical compounds are made up of two elements in set proportions to each other. Water provides an easy example of a compound. It is composed of the elements hydrogen and oxygen. There are always two parts of hydrogen to one part oxygen in every molecule of water. If the water sample is from the sea or polluted, there may be other chemicals added, but basic water always has the same proportion of hydrogen to oxygen by mass.

Percent

In order to better understand how elements combine to become different kinds of matter, it is important to understand the idea of *percent*. The word, cent, comes from the Latin word *centum* which means one hundred. Percent stands for the number of parts of one material included in the total amount of another sample.

Scientists in all areas of study use the concept of percent to do their analyses. The general formula for percent looks like the following:

$$x/x_{\text{total}} = n/100$$

Being able to calculate percent easily will make your laboratory experiments a lot less likely to give you problems. By starting out with the correct percentages and ratios of reactants, your chances of success (or at least a good grade) increase dramatically. Try some of the following examples.

EXAMPLE 3.1

Consider a rancher who has 100 horses: 24 are Appaloosas, 16 are Bays, 20 are Paints, and 40 are Palominos. What percent of Appaloosas does he have? He has 24 out of a hundred or 24%.

$$24/100 = n/100$$

$$n = 24\%$$

EXAMPLE 3.2

On a different scale, consider a typical classroom of 30 people. In this group, 3 are interested in English and journalism, 12 are interested in medicine and science, 6 are interested in teaching and education, and the final 9 want to pursue art, communications, and film. Since 30 is the total number of people in the class, we treat them as the whole or 100%. To figure out the percent of people who want to pursue various areas of further study, you divide by 30.

$$3/30 = 0.1 \qquad 0.1 \times 100 = 10\%$$

(Multiplying by 100 allows you to figure out what percent of the whole is given.) 10% of the people in the class want to study English and journalism.

$$12/30 = 0.4 \qquad 0.4 \times 100 = 40\% \text{ (science and math)}$$

$$6/30 = 0.2 \qquad 0.2 \times 100 = 20\% \text{ (teaching and education)}$$

$$9/30 = 0.3 \qquad 0.3 \times 100 = 30\% \text{ (art, communications, and film)}$$

In chemistry, percent is used to record the amounts of an element within the entire material or form.

Let's look at the human body. It is composed of many different compounds in various amounts. In science fiction terms, humans are referred to as "carbon-based units" since the average human is composed of 18% carbon by mass out of all the elements in the body. **Figure 3.4** gives an idea of the variety of elements that combine to keep our bodies healthy.

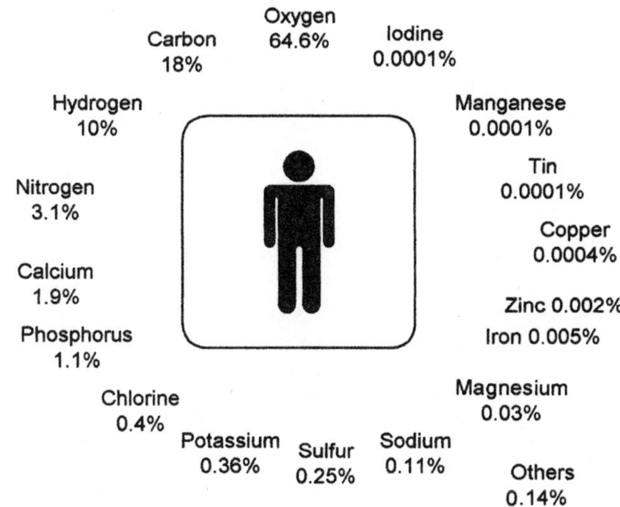

Fig. 3.4. A wide variety of elements are found in the body.

On a larger scale, the Earth is composed of many of the same elements found in our bodies as well as additional minerals and metals. **Figure 3.5** lists the more plentiful elements in the Earth's surface. The diversity of these elements supports and provides for millions of species of plants, animals, and minerals found on this planet. Without this large resource of diverse elements, or if it should swing far in the direction of one element over the others, our planet could become as barren as many others in our solar system.

One of the interesting things about our planet is the wide distribution of elements around the globe. Some of this diversity is due to the original formation of the continents and some is due to environmental factors, but the elements' diverse properties play a huge role in making our world what it is today.

Without many different elements, life on Earth might still be just primordial ooze floating in pockets of sludge upon a primitive sea.

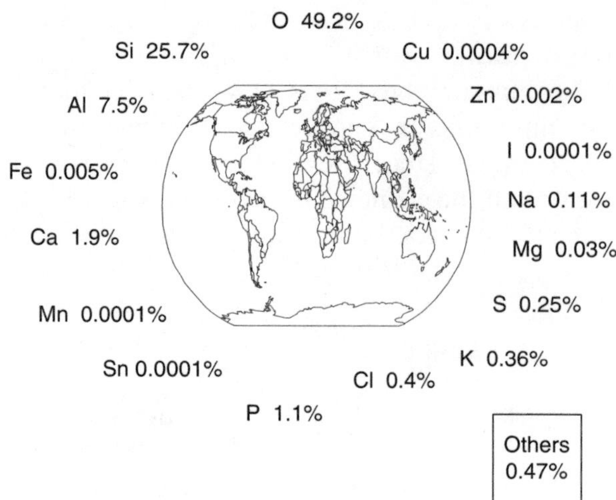

Fig. 3.5. Elements of the Earth are varied, but many are very low in quantity.

We will learn more about the variety of elements, their specific characteristics, and their groupings in Chapter 4.

Quiz 3

1. Matter is
   (a) a legal term
   (b) something found in the lower atmosphere
   (c) anything that has mass and occupies space
   (d) anything that can be seen

2. Atoms
   (a) are found only in solid materials
   (b) were thought to be next to the smallest particles
   (c) were first found in ants
   (d) contain particles such as protons and neutrons

3. Physical properties of matter
   (a) contain all heat-bearing compounds
   (b) include color, form, density, and boiling point
   (c) are only visible under ultraviolet light
   (d) are directly related to muscle mass

4. Pure substances
   (a) are homogeneous and have unchanging chemical compositions
   (b) are heterogeneous and have unchanging chemical compositions
   (c) seldom exist in nature
   (d) are found in cubic form only

5. Antoine Lavoisier
   (a) is called the father of modern chemistry
   (b) identified 33 elements
   (c) should have stayed away from government taxation
   (d) all of the above

6. Solids
   (a) have a measurable volume and can change shape
   (b) are always heavy to transport
   (c) are fixed and rigid with a measurable volume
   (d) change shape with little effort

7. Chemical properties
   (a) describe a material's behavior when acted on by something else
   (b) are those things that can be seen
   (c) are associated with water's freezing point
   (d) are defined as solids, liquids, and gases

8. An element
   (a) is animal, vegetable, or mineral
   (b) is made up of a pure chemical and not divided into simpler parts
   (c) can be separated into neutrons and protons
   (d) is a six-sided solid

9. Percent
   (a) comes from the Latin word for machine
   (b) can be calculated for diamond facets
   (c) is the number of parts of one material found in another
   (d) is an ancient form of Greek counting

10. Barium is
    (a) green in color and melts at 627°C
    (b) never used in medicine
    (c) a different compound completely when melted
    (d) silvery white and found in the solid state

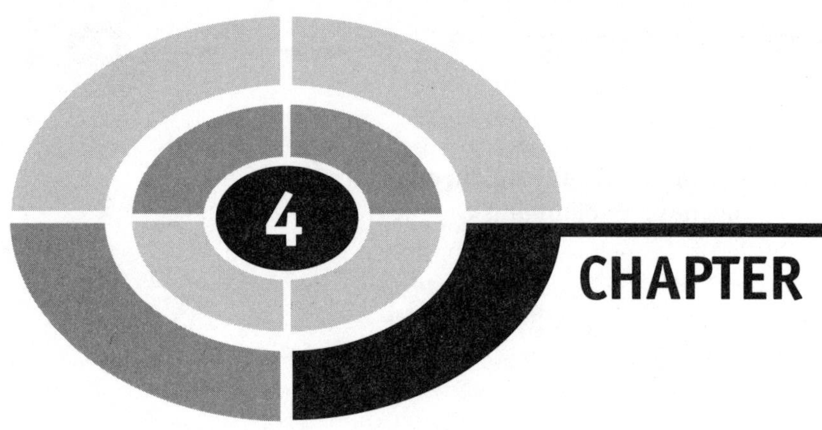

Elements, Symbols, and the Periodic Table

Chemical Nomenclature

Before standard element symbols and the Periodic Table were invented, discussing matter was a guessing game. Early scientists located in laboratories far distant from each other and speaking different languages had problems communicating. Cultural and language differences had a big impact on the naming of elements.

Additionally, just as ancient words and phrases are no longer part of today's spoken languages, like "thee" and "thou" of old English, some element names originated from seldom used languages. The ancient name for copper is *cuprum*, which is why the symbol for copper is Cu. Scientists, confused by outdated names, needed a standard method that everyone used.

Nicknames added to the confusion. For example, baking soda used to make breads rise is actually sodium bicarbonate, but most people don't call it that. Battery acid, the liquid that allows electricity to be mysteriously generated in cars, is really sulfuric acid. Household laundry bleach is usually known as sodium hypochlorite by chemists. Or do you know anyone who says, "Please pass the sodium chloride," at a meal?

Common names for elements don't provide accurate descriptions of elemental components. In fact, there are so many compounds and combinations of chemicals that sometimes figuring out what someone is talking about takes longer than the experiment!

Elements

For thousands of years, people have known about the basic elements. They knew rocks were hard, water was liquid, and fog was a mist. They knew materials could be heated, packed down, frozen, and altered in different ways, while others could not. What they didn't understand was how matter seemed to turn from one form into another. Rust was a mystery. Explanations, based more on ancient myth than science, provided a path to understanding.

Symbols

As time went on, scientists began to concentrate on the study of individual elements, but they still had a problem. Because in different languages chemical elements were known by different names, scientists couldn't always be sure they were talking about the same thing. Just as a traveler, not knowing the language, has problems in foreign countries asking directions or finding a hotel, early chemists had problems comparing results and analyzing compounds that no one seemed to recognize. For example, the element, iron, is called Eisen in German, Piombo in Italian, Olovo in Czech, and Fer in French. Scientists, too busy with their experiments to study languages, ran into trouble when they tried to communicate their findings to colleagues in other parts of the world. To give you an idea of the problem, **Table 4.1** provides a sampling of element names in different languages.

Table 4.1 Elements have many interesting names in different languages.

Symbol	English	German	Swedish	French	Spanish
Au	Gold	Gold	Guld	Or	Oro
Bi	Bismuth	Wismut	Vismut	Bismuth	Bismuto
C	Carbon	Kolenlstoff	Kol	Carbone	Carbono
Co	Cobalt	Kolbalt	Kolbolt	Cobalt	Cobalto
Cu	Copper	Kupfer	Koppar	Cuivre	Cobre
Fe	Iron	Eisen	Jarn	Fer	Hierro
Hg	Mercury	Quecksilber	Kvicksilver	Mercure	Mercurio
N	Nitrogen	Stickstoff	Kvave	Azote	Nitrogeno
Na	Sodium	Natrium	Natrium	Sodium	Sodio
P	Phosphorus	Phosphor	Fosfor	Phosphore	Fosforo
Pb	Lead	Blei	Bly	Plomb	Plomo
S	Sulfur	Schwefel	Svavel	Soufre	Azufre
Se	Selenium	Selen	Selen	Selenium	Selenio
Sn	Tin	Zinn	Tenn	Etain	Estano
W	Tungsten	Wolfram	Volfram	Tungstene	Wolframio
Z	Zinc	Zink	Zink	Zinc	Zinc

It was obvious that some sort of common code or *chemical nomenclature* was needed.

Chemical nomenclature is the standardized system used to name chemical compounds.

Symbols, at first based on Latin words, were used as an elemental code because writing the full name was time consuming. The powerful insecticide dichlorodiphenyltrichloroethane, or DDT, is written as $(C_6H_4Cl)_2CHCCl_3$ with Berzelius' method. If it weren't for this shorthand code so much time would be taken up writing the samples' names that there wouldn't be any time left to do the experiment!

Chemical shorthand becomes especially important when writing chemical reactions as will be seen in later chapters. The symbol for an element can be one letter as in carbon (C) and phosphorus (P), two letters as in strontium (Sr) and molybdenum (Mo), or three letters as in the more recent elements in the Periodic Table such as ununquadium (Uuq) and ununoctium (Uuo). Notice that when an element has more than a one letter shorthand name, only the first letter of the symbol name is capitalized.

In 1862, the French geologist Antoine Béguyer de Chancourtois made up a list of elements, arranged by increasing atomic weight. He is said to have wrapped the list, divided into 16 sections around a cylinder. After he had done this, he noticed that different sets of similar elements lined up. One of these groups, oxygen, sulfur, selenium, and tellurium, had a repeated pattern. The atomic weights of these elements are 16, 32, 79, and 128, all multiples of 16. This periodic repeat seemed to be part of a natural pattern that occurred regularly.

Atomic weight is not to be confused with *atomic number*. Atomic number is written as the superscript of an element on the Periodic Table, while the atomic weight is written as a subscript. The atomic weight of boron is 10.81 while its atomic number is 5.

Atomic number (Z) equals the number of protons in the nucleus of an atom.

Around this same time, English chemist John Newlands, Professor of Chemistry at the School of Medicine for Women in London, was also overlapping elements and seeing similarities. With a repeat of chemical groups every eight elements, Newlands, a jazzy kind of guy, was reminded of eight-note music intervals and called his findings the *Octave rule*. Members of the Royal Chemical Society, not into music composition, ignored Newlands' work for many years.

In 1864, *Die Modernen Theorien der Chemie (The Modern Theory of Chemistry)* was published by German chemist Lothar Meyer. Meyer used the atomic weight of elements to arrange twenty-eight elements into six families with similar chemical and physical characteristics. Sometimes elements seemed to skip a predicted weight. Where he had questions, he left spaces for possible elements. Meyer also used the word *valence* to describe the number that equals the combining power of one element with the atoms of another element. He thought this combination of characteristic grouping and valence was responsible for the connection between families of elements.

After seeing a recurring pattern of peaks and valleys when plotting atomic weight, Meyer thought these patterns seemed to make up family rhythms. When he measured the volume of one atomic weight's worth of an element with the same number of atoms in the sample, Meyer decided the measurements must stand for the same amounts of each distinct atom.

Figure 4.1 illustrates Meyer's experimental results with a recurring pattern in one family. If you start with the element at the top of each peak, you find lithium, sodium, potassium, rubidium, and cesium lined up by atomic number and weight.

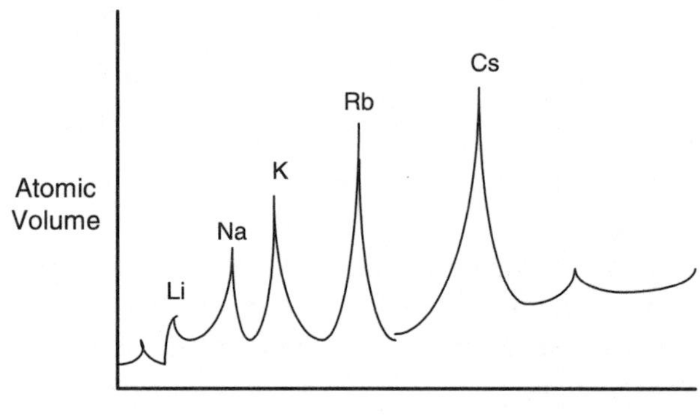

Fig. 4.1. Meyer's research results were divided by atomic weight.

Unlike Newlands' octaves, Meyer's data showed the groups were not the same length. He was one of the first chemists to notice that the group lengths changed. Hydrogen was in a group by itself, lithium through fluorine made another group, sodium through chlorine made another, potassium through bromine, rubium through iodine, and others. The groups started out small and then became larger.

Meyer saw repeating periods of atomic volume, but they changed in size. The first period contained only hydrogen and was one period in length. The second and third periods had seven elements. The fourth and fifth periods were seventeen elements in length. Meyer's work held until the inert gases were discovered, then an extra element was added to each period, to give periods of 2, 8, 8, 18, and 18.

Five years later, working separately, Russian chemist Dimitri Mendeleyev and Meyer arranged the elements into seven columns relative to the elements' known physical and chemical properties. The differences in their work were slight, but each added to the current knowledge.

Mendeleyev presented a scientifically significant table of elements in his paper *On the Relation of the Properties to the Atomic Weights of the Elements*, which the Russian Chemical Society praised. In his paper, Mendeleyev discussed the periodic resemblance between chemical groups with respect to similar reactions.

As it turns out, many of the gaps Mendeleyev made spaces for in his periodic chart turned out to be correct placeholders for elements discovered later. Initially, titanium (Ti) was placed next to calcium (Ca), but this would have placed it in group III with aluminum (Al). However, since studies of titanium's properties showed that it was more like silicon (Si) than aluminum, Mendeleyev left a space next to calcium and eventually placed Ti in group IV with silicon.

Ten years later in 1879, Swedish chemist Lars Nilson discovered the missing element with properties in between calcium (atomic weight of 40) and titanium (atomic weight of 48) and named it scandium (Sc). Scandium has an atomic weight of 45.

Figure 4.2 shows some of the characteristics of the element titanium.

Ti

Titanium

Atomic Number – 22

Atomic Mass – 47.90

Group – 4

Period – 4

Transition Metal

Electrons per orbital layer – 2,8,10,2

Valence electrons – 1s2 2s2p6 3s2p6d2 4s2

Fig. 4.2. Specifics of the metal titanium include its group and period.

In 1870, Meyer's next generation Periodic Table of 57 elements was published. This table, including properties such as *melting point*, added depth to the understanding of interactions and the role of atomic weight at that time.

Meyer additionally studied the atomic volume of elements to fine tune his placement of elements into particular groups.

Perhaps Meyer's curiosity came from the fact that he grew up in a family of physicians and was exposed to scientific and medical discussions for much of his early life. His initial schooling in Switzerland was in the field of medicine. The many elements in the body and their complex interactions gave Meyer much to think about. **Table 4.2** shows a few of the most common elements and their functions in the body.

Table 4.2 Elements serve many important functions in the body.

Element	**Functions in the body**
Calcium	Bones, teeth, and body fluids
Phosphorus	Bones and teeth
Magnesium	Bone and body fluids, energy
Sodium	Cellular fluids, transmission of nerve impulses
Chloride	Dissolved salt in extracellular and stomach fluids
Potassium	Cellular fluids and transmission of nerve impulses
Sulfur	Amino acids and proteins
Iron	Blood hemoglobin, muscles, and stored in organs

Meyer's initial chemistry research grew out of his fascination with the physiology of respiration. He was one of the first scientists to recognize that oxygen combined with hemoglobin in the blood. Then, in order to explain specific biochemical processes and systems, he found he had to identify the elements more completely.

The modern Periodic Table contains around 118 elements. Those up through atomic number 92 (uranium) are naturally occurring, whereas the "transuranic" elements, those synthesized in heavy-nuclei interactions, make up the most recent discoveries. Some new elements' symbols, like in the time of Meyer and Mendeleyev, represent gaps or spaces for elements that seem to be hinted at by test data. When compared to Mendeleyev's and Meyer's early tables, the details described over 150 years ago are amazingly accurate. The modern Periodic Table is shown in **Figure 4.3** and on the inside cover of this book.

More information on individual elements can be found on the Internet at a variety of sites including:

http://www.webelements.com

http://environmentalchemistry.com/yogi/periodic/Pb.html

Periodic Table of Elements

Periodic Table History

	I A	II A	III A	IV A	V A	VI A	VII A	VIII A	VIII A	VIII A	I B	II B	III B	IV B	V B	VI B	VII B	VIII B
1	1 H 1.0079																	2 He 4.003
2	3 Li 6.94	4 Be 9.0121											5 B 10.81	6 C 12.011	7 N 14.006	8 O 15.999	9 F 18.998	10 Ne 20.17
3	11 Na 22.989	12 Mg 24.035											13 Al 26.981	14 Si 28.085	15 P 30.973	16 S 32.06	17 Cl 35.453	18 Ar 39.948
4	19 K 39.098	20 Ca 40.08	21 Sc 44.955	22 Ti 47.90	23 V 50.941	24 Cr 51.996	25 Mn 54.938	26 Fe 55.847	27 Co 58.933	28 Ni 58.71	29 Cu 63.546	30 Zn 65.38	31 Ga 69.735	32 Ge 72.59	33 As 74.921	34 Se 78.96	35 Br 79.904	36 Kr 83.80
5	37 Rb 85.467	38 Sr 87.62	39 Y 88.905	40 Zr 91.22	41 Nb 92.906	42 Mo 95.94	43 Tc 98.906	44 Ru 101.07	45 Rh 102.90	46 Pd 106.4	47 Ag 107.86	48 Cd 112.41	49 In 114.82	50 Sn 118.69	51 Sb 121.75	52 Te 127.60	53 I 126.90	54 Xe 131.30
6	55 Cs 132.90	56 Ba 137.33	57 La 138.90	72 Hf 178.49	73 Ta 180.94	74 W 183.85	75 Re 186.20	76 Os 190.2	77 Ir 192.22	78 Pt 195.09	79 Au 196.96	80 Hg 200.59	81 Tl 204.37	82 Pb 207.2	83 Bi 208.98	84 Po (209)	85 At (210)	86 Rn (222)
7	87 Fr (223)	88 Ra 226.02	89 Ac (227)	104 Unq (261)	105 Unp (262)	106 Unh (263)	107 Uns (262)	108 Uno (265)	109 Une (266)	110 Unn (272)								
Lanthanide Series			58 Ce 140.12	59 Pr 140.90	60 Nd 144.24	61 Pm (145)	62 Sm 150.4	63 Eu 151.96	64 Gd 157.25	65 Tb 158.92	66 Dy 162.5	67 Ho 164.93	68 Er 167.26	69 Tm 168.93	70 Yb 173.04	71 Lu 174.96		
Actinide Series			90 Th 232.03	91 Pa 231.03	92 U 238.02	93 Np 237.04	94 Pu (244)	95 Am (243)	96 Cm (247)	97 Bk (247)	98 Cf (251)	99 Es (254)	100 Fm (257)	101 Md (258)	102 No (259)	103 Lr (260)		

Color Reference for element type:	
Noble Gas	Halogen
Metal	Rare Earth
Trans. Metal	Non Metal
Alkali Metal	Alkali Earth

Fig. 4.3. The Periodic Table is an important tool for the chemist.

Periods and Groups

Like a family history, the elements are arranged in family groups such as noble gas, halogen, metal, rare earth, transitional metal, non-metal, alkali metal, and alkaline earth. Just as genetic analysis helps biologists and physicians to determine a person's make-up, so the grouping of elements into families and groups helps chemists to understand similar properties of different elements.

Elements with similar boiling or melting points usually act the same way when exposed to the same experimental conditions. The same is true of freezing and vaporization points.

The Periodic Table is the most important tool in general chemistry. Probably only the Bunsen burner (laboratory gas flame) rivals it in a distant second place. The amount of information pulled together in one place makes calculations, reactions, and the study of matter a whole lot easier to decipher.

Basically, the Periodic Table is divided into rows and columns, known as *periods* and *groups*. Dividing elements into periods and groups helps classify them by their specific characteristics.

PERIODS

Each period ends with an element known as a noble gas. Like kings and queens set apart in impenetrable castles, these gases are chemically unreactive and composed of individual atoms.

> A **period** contains the elements in one horizontal row of the Periodic Table.

The first period contains only hydrogen (H) and helium (He). The second period has 8 elements: lithium (Li) through neon (Ne). The third period also contains 8 elements: sodium (Na) through argon (Ar). The fourth period has 18 elements: potassium (K) through krypton (Kr). The fifth period also has 18 elements: rubidium (Rb) through xenon (Xe). The sixth period has 32 elements; cesium (Cs) through radon (Rn). To make the table less bulky, the sixth and seventh period rows have been divided between 57, 58 and 89, and 90. These rows are shown fully expanded at the bottom of the chart. The seventh period is not complete, but includes gaps just as Meyer did earlier to allow for additional elements: francium (Fr) through lawrencium (Lr). Placeholders are in debate through element 118 and sometimes are not included on older Periodic Tables.

GROUPS

The groups of the Periodic Table are numbered most frequently with Roman numerals. The International Union of Pure and Applied Chemistry (IUPAC), in order to avoid confusion, set up a standard numbering plan in which columns were numbered I–VIII, according to their characteristics.

> A **group** contains the elements in one column of the Periodic Table.

These groups are further divided into A and B sub-groups with the A groups called the main groups or *representative* elements and the B groups called the *transition* elements. Numbers 58 (cerium) to 71 (lutetium) are known as the *lanthanide* series and 90 (thorium) to 118+ (ununoctium) as the *actinide* series of elements.

The Periodic Table divided into periods and groups is shown in **Figure 4.4**.

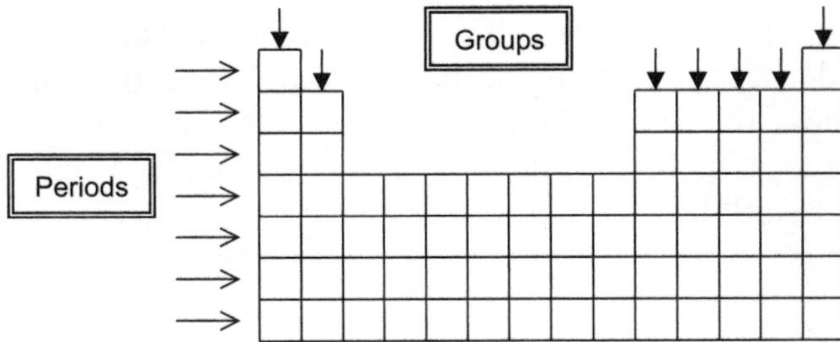

Fig. 4.4. Groups and periods of elements are found in columns and rows, respectively, on the Periodic Table.

Element Classes

As Meyer reported in his research, some elements with similar properties can be grouped together. Basically, groups of elements are divided into four main classes.

(1) Representative elements (groups IA–VIIA)
(2) Noble or inert gases (group VIII)

(3) Transition metals (group B elements)
(4) Inner transition metals

The *representative elements* or main group of elements is further defined. Group IA, the *alkali metals* (e.g., lithium, potassium), are all soft metals (except hydrogen, a gas) that react readily with water. Group IIA, known as the *alkaline earth metals* (e.g., beryllium, cadmium), are also reactive chemically. The *halogens*, in group VIIA, are all non-metals. *Chalcogens*, group VIA (e.g., oxygen, sulfur), comes from the Greek word *chalkos* meaning ore. Many ores are made with varying amounts of oxygen and sulfur. The remainder of the representative element groups (IIIA–VA) have not been given descriptive names.

The *noble* or *inert gases* (group VIII) are called inert since they seldom form chemical compounds. In fact, helium, neon, and argon refuse to play with anyone and don't form any compounds at all. All these gases exist naturally as individual atoms in the environment.

The *transition metals* (group B) contain the more recognizable metals. They are used in construction, coins, and jewelry. The transition metals group includes iron, nickel, and chromium as well as gold, silver, and copper.

The *inner transition metals* consist of the 15 *rare earth* metals or lanthanides. They are all silvery white in color and used in such products as permanent magnets and headphones. The other inner transition metals, a set of elements named after the element actinium, include uranium, americium, and neptunium. They are primarily human-made elements. These metals are radioactive and used in advanced smoke detectors, neutron-detection devices and in nuclear reactions.

Metals vs. Non-metals

It is important to observe that metals and non-metals are shown on the Periodic Table by a heavy zigzag line with metals to the left side and non-metals to the right. **Figure 4.5** shows this dividing line.

Most **metals** are shiny and good conductors of heat and electricity.

Metals, about 80% of the elements, can be pulled into thin wires (ductile) or pounded into sheets (malleable). Mercury is the only metallic element that is liquid at room temperature.

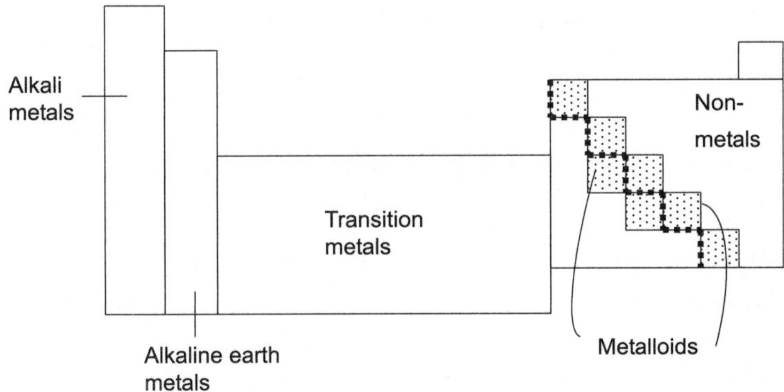

Fig. 4.5. There are a lot more metals than non-metals.

Non-metals are basically everything else. Most are gases such as helium and argon, or brittle solids such as phosphorus and selenium. Bromine is the only liquid, non-metallic element at room temperature.

The elements found along the borderline of metals and non-metals are known as *semi-metals* or *metalloids* since they have the characteristics of both metals and non-metals. For example, silicon is used to make lubricants, computer circuits, and medical implants and joints.

We will take a closer look at metals and non-metals in Chapter 12 and discover why many elements are placed where they are in the Periodic Table.

If asked to learn only one thing in all of chemistry, pick the Periodic Table. Learn it and all the rest will fall into place.

Quiz 4

1. When John Newlands made a list of the elements in the 1860s, he
   (a) noticed the list seemed far longer than he remembered
   (b) wrapped the list around his favorite mineral sample
   (c) saw that the elements lined up and repeated in groups of eight
   (d) saw similarities between elements and called it the Newlands rule

2. Who published *Die Modernen Theorien der Chemie* in 1864?
   (a) Johnnes Kepler
   (b) Lothar Meyer
   (c) Antoine Beguyer de Chancourtois
   (d) Dimitri Mendeleyev

3. Currently, the modern Periodic Table contains how many elements?
   (a) 57
   (b) 88
   (c) 109
   (d) between 112 and 118 depending on which research papers you read

4. Which of the following is not an elemental family group?
   (a) halogen
   (b) rare Earth
   (c) alkali Earth
   (d) alloy

5. The following are all names for the element sulfur, except
   (a) schwefel
   (b) svavel
   (c) selenur
   (d) azufre

6. Over 100 years ago, the Periodic Table contained gaps because
   (a) experimental data hinted at elements in between known elements
   (b) researchers couldn't agree on which elements to include
   (c) experimental equipment wasn't accurate enough to find elements
   (d) scientists didn't transcribe earlier Periodic Tables correctly

7. The formula $(C_6H_4Cl)_2CHCCl_3$ is shorthand for what compound?
   (a) acetyl chloride
   (b) dichlorodiphenyltrichloroethane
   (c) polypropylene dichloride
   (d) chlorofluorocarbon

8. The standardized system of naming chemical compounds is called
   (a) the Bernouli rule
   (b) the Periodic Table
   (c) chemical nomenclature
   (d) the Octave rule

9. All the metals listed below are solid at room temperature except
   (a) mercury
   (b) tin
   (c) molybdenum
   (d) iron

10. The 18 elements potassium (K) through krypton (Kr) are found in
    (a) group 7 of the Periodic Table
    (b) period 6 of the Periodic Table
    (c) group 5 of the Periodic Table
    (d) period 4 of the Periodic Table

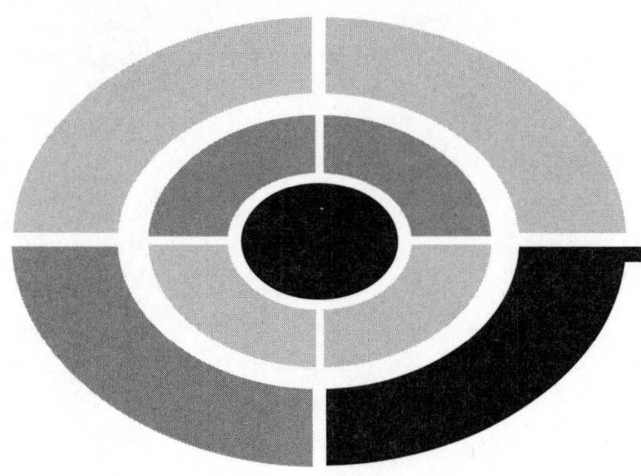

Test: Part One

1. Which of the following does not describe a physical property?
   (a) silvery color
   (b) formation of rust
   (c) ice melting
   (d) erosion

2. Early chemists wanted to
   (a) turn lead into gold
   (b) figure out why iron rusted
   (c) fashion tools and bowls from metal
   (d) all of the above

3. A statement or idea that attempts to explain observed information is called a
   (a) scheme
   (b) fact
   (c) hypothesis
   (d) mission statement

4. Early chemists who wanted to turn lead to gold were called
   (a) politicians
   (b) carpet baggers

(c) crazy
(d) alchemists

5. To perform this activity, you must have a control, a sample to be tested, and make careful observations and measurements
(a) scuba diving
(b) experiment
(c) yoga
(d) evaporation

6. Antoine Lavoisier is called the
(a) father of modern chemistry
(b) discoverer of beryllium
(c) father of quark theory
(d) brother of Larry Lavoisier

7. This idea predicts the results of testing based on past experimental data
(a) imagination
(b) hunch
(c) theory
(d) rumor

8. The atomic theory was first described by
(a) Albert Einstein
(b) Albert Schweitzer
(c) John Dalton
(d) Leonardo da Vinci

9. What is the difference between a scientific theory and a law?
(a) the president has to sign a bill into law
(b) a theory can be tested, while a law is a hair-brained idea
(c) you are not usually arrested for performing a theory
(d) a scientific law is tested many times and believed to be without exception, while a theory is an idea

10. The total partial pressure measurement of several combined gases is achieved by
(a) using a logarithmic calculator
(b) adding together the individual pressures of each gas
(c) adding together the first two gases and dividing by the third
(d) multiplying the individual pressures of each gas

11. The difference between alchemists' work and that of today's chemists is
 (a) the types of machinery used
 (b) science today is based on many repeated experiments by chemists all over the world
 (c) that alchemy was often performed to get rich
 (d) slim to none

12. A decimal system of recording data was first described by
 (a) Benjamin Franklin
 (b) Antoine Lavoisier
 (c) Gabriel Mouton
 (d) Plato

13. The International Bureau of Weights and Standards uses what metal as a standard?
 (a) gold
 (b) silver
 (c) titanium
 (d) platinum

14. Chemistry is an experimental science divided into
 (a) pure chemistry and applied chemistry
 (b) acids and bases
 (c) protons and electrons
 (d) quarks and mesons

15. A method to write numbers in powers of 10 is called
 (a) rounding
 (b) exponential notation
 (c) includes writing all zeros in the number out
 (d) mathematical notation

16. 1/10,000 is written as what in exponential notation?
 (a) 10^{-2}
 (b) 10^{-4}
 (c) 10^{-5}
 (d) 10^{-6}

17. A single measurement closest to its true value is the most
 (a) reliable
 (b) tested
 (c) accurate
 (d) precise

18. Rounding is commonly used to
    (a) cook pizza
    (b) number paint-by-number kits
    (c) win at horseshoes
    (d) number whole objects, like chickens

19. Dimensional analysis
    (a) compares squares to triangles
    (b) studies the number of molecules in the ocean
    (c) is a method of studying carbons to oxygen in the atmosphere
    (d) changes one unit to another by using conversion factors

20. To measure absolute zero, you use the
    (a) Kelvin scale
    (b) Fahrenheit scale
    (c) Celsius scale
    (d) bathroom scale

21. Antoine Lavoisier
    (a) shouldn't have gotten involved with French taxation
    (b) insisted on precise measurements
    (c) described the properties of matter
    (d) all of the above

22. Subatomic particles are
    (a) found in the core of an atom's nucleus
    (b) only found on nuclear submarines
    (c) made up of about 10^3 millimeters in length
    (d) found on big sandwiches

23. Which of the following commonly exists as a solid, liquid, and gas?
    (a) carbon dioxide
    (b) nitrous oxide
    (c) water
    (d) hydrogen

24. Which form of matter is bendable, takes the shape of its container, and is pourable?
    (a) gas
    (b) solid
    (c) crystal
    (d) liquid

25. An element has how many classes of properties to describe it?
    (a) 1
    (b) 2
    (c) 3
    (d) 4

26. When copper turns green, it is an example of its
    (a) chemical property
    (b) physical property
    (c) value
    (d) thickness

27. Lavoisier identified how many elements that he thought were pure and indivisible?
    (a) 18
    (b) 27
    (c) 33
    (d) 42

28. A chemical experiment is
    (a) never done in the laboratory
    (b) a carefully controlled and measured testing of a sample's properties
    (c) something to try once and then move on to other things
    (d) always performed at room temperature

29. Which of the following samples is not a single, pure element in nature?
    (a) oxygen
    (b) mercury
    (c) nickel
    (d) iron

30. Which of the following is not a physical property of gold?
    (a) it is highly reactive
    (b) it melts at 30°C
    (c) it is naturally found as a solid
    (d) it has a luster

31. The biggest problem scientists had before the Periodic Table was invented was
    (a) dental hygiene
    (b) cultural and language differences
    (c) funding
    (d) finding time for a social life

32. Battery acid is the common name for
    (a) formic acid
    (b) hydrochloric acid
    (c) nitric acid
    (d) sulfuric acid

33. Who liked music and came up with the octave rule?
    (a) Antoine Béguyer de Chancourtois
    (b) Charles Darwin
    (c) John Newlands
    (d) Amadeus Mozart

34. What was the element discovered between titanium and calcium?
    (a) silicon
    (b) scandium
    (c) lead
    (d) potassium

35. The chemical shorthand name for molybdenum is
    (a) Mo
    (b) Mb
    (c) Md
    (d) Mn

36. Plomb is the French name for which element?
    (a) lead
    (b) iron
    (c) platinum
    (d) potassium

37. How many carbons are in dichlorodiphenyltrichloroethane?
    (a) 10
    (b) 12
    (c) 14
    (d) 16

38. Meyer saw that element groups repeated, but was the first to notice that
    (a) the groups were not all of the same length
    (b) the groups were all eight elements long
    (c) the overlap was not seen in the alkaline earth metals
    (d) the groups were all eighteen elements long

39. Tera is the prefix used to show
    (a) 10^2
    (b) 10^6
    (c) 10^9
    (d) 10^{12}

40. Lothar Meyer recognized that hemoglobin combined with what element in the blood?
    (a) helium
    (b) mercury
    (c) nitrogen
    (d) oxygen

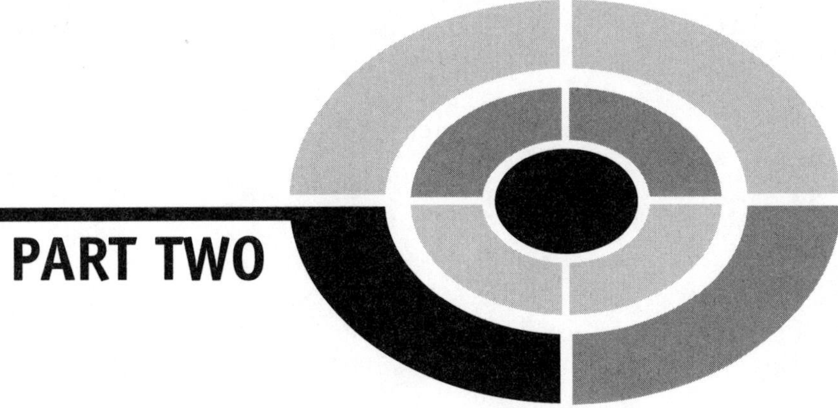

PART TWO

Chemical Building Blocks

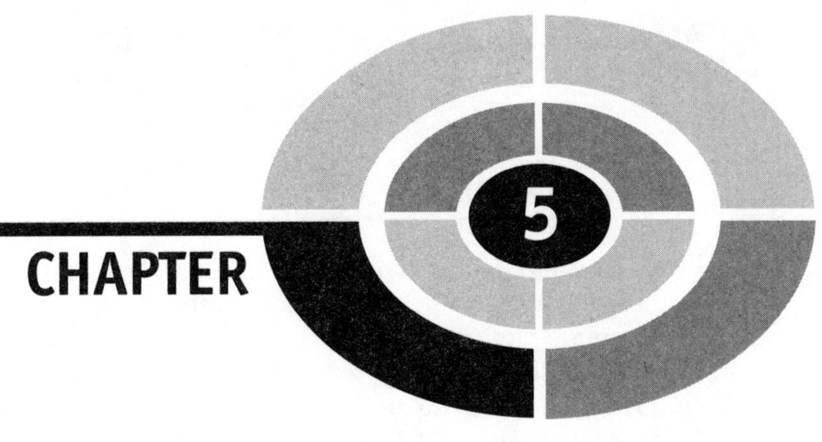

Atoms, Elements, and Compounds

Electrons

In 1897, J.J. Thomson, a physicist from England, discovered negatively charged particles by removing all the air from a glass tube that was connected to two electrodes. One electrode was attached to one end of the tube and negatively (–) charged. It was called the *cathode*. The other end of the tube was attached to a positive (+) electrode and called the *anode*. A *cathode ray tube* uses a current to excite atoms of different gases contained in the tube. The electricity is beamed directionally through the length of the tube to the other electrode. By using this piece of equipment, scientists of a century ago began to separate the individual particles of atoms.

Through his experiments with several different colored gases, Thomson found that *electrons* had a negative charge and seemed to be common to all elements.

> **Electrons** are small negatively charged sub-atomic particles that orbit around an atom's positively charged nucleus.

However, since Thomson's results showed that the overall charge of atoms was neutral in nature, something within the atom must be positive to counteract the negative charge. This something made the atom neutral.

Thomson came up with the "plum pudding" model of sub-particle arrangement made up of a blob of positively (+) charged particles, the pudding, and specks of negatively (–) charged particles floating around in it like raisins. He probably ate dessert right before or after working in his lab, so the idea came to him fairly easily. The plum pudding model of electrons and protons is shown in **Figure 5.1.**

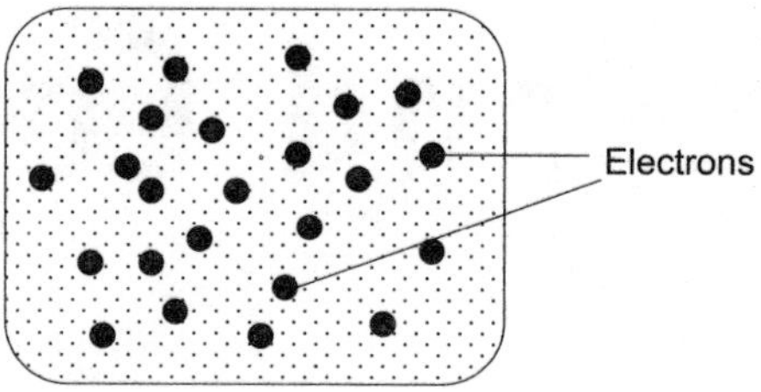

Fig. 5.1. The plum pudding model of electrons and protons was not compact.

In 1906, Thomson was awarded the Nobel Prize for physics for his research and electrical work with gases. Later research found that an electron has a mass of 9.1×10^{-31} kg and that it has a charge of 1.6×10^{-19} Coulombs.

It wasn't until a student of Thomson's, Ernest Rutherford, started working to support his teacher's ideas that the data for a plum pudding model just didn't hold up. The floating negatively charged "raisins" acted differently in electrical current, for different elements, than what Thomson expected. This seemed to suggest they had different energy levels. (Maybe that is where the expression "the proof is in the pudding" came from.)

The Nucleus

It wasn't until scientists discovered that the atom was not just a solid core, but made up of smaller building block sub-particles located in the nucleus, that some of their data made sense.

In 1907, Rutherford, teaching at Cambridge, developed the modern atomic concept. He received the Nobel Prize for Chemistry in 1908 and was knighted in 1914 for his work. (Whoever said chemistry was not a glory science?)

Through his experiments with radioactive uranium in 1911, Rutherford described a *nuclear* model. By bombarding particles through thin gold foil, he predicted that atoms had positive cores that were much smaller than the rest of the atom.

Instead of thinking that atoms were the same all the way through ("plum pudding" model) as Thomson suggested, Rutherford's experiments pointed more toward something like a fruit with a small, dense pit. His experiments along with those of his student, Hans Geiger, showed that over 99% of the bombarded particles passed easily through the gold, but a few (one out of eight thousand) ricocheted at wild angles, even backwards. **Figure 5.2** shows how Rutherford's dense pit model of the nucleus might look.

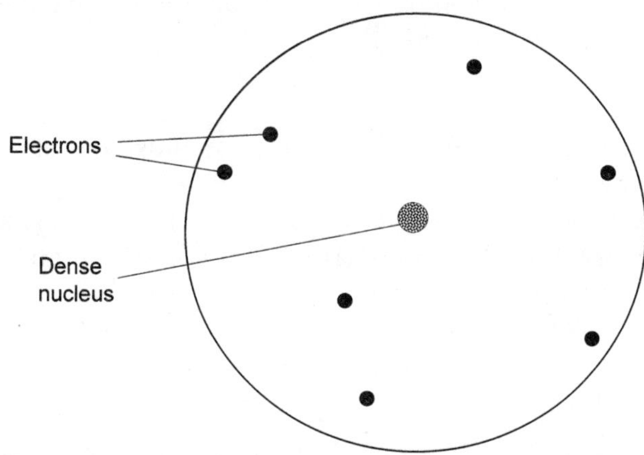

Fig. 5.2. Rutherford's model of the nucleus had a tight central core.

Rutherford thought this scattering happened when positive nuclei of the test particles collided and were then repelled by heavy positively charged gold nuclei. It was later proven that Rutherford's dense pit model was correct. When an accelerated alpha particle collided with an electron of a gold atom in a gas, a proton was knocked out of the nucleus.

Later research done along the same lines as Rutherford's early work found that protons in a nucleus have a mass over 1800 times that of an electron. In fact, the positively charged nucleus of the atom that contained most of its mass was very dense and took up only a tiny part of an atom's total space.

To get an idea of size, if an atomic nucleus were the size of a ping-pong ball, then the rest of the atom with its encircling negatively charged electrons would measure nearly 3 miles across. More precisely, nuclei are roughly 10^{-12} meters in diameter. The total diameter of an atom is around 10^{-8} meters or roughly 10,000 times larger.

PROTON

A proton has a positive charge and roughly 1800 times greater mass than an electron.

> A **proton** is a smaller bit of positively charged matter or **sub-atomic particle** within the nucleus.

The *atomic number* (Z) of an element is taken from the number of protons in the nucleus of an atom. A pure element is one that is made up of particles that all have the same atomic number.

EXAMPLE 5.1

To obtain the atomic number of an element, you must identify the number of protons in the nucleus.

What is the atomic number (Z) of (a) boron, (b) gold, (c) zinc, (d) iridium, and (e) bismuth? Did you get (a) 5, (b) 79, (c) 30, (d) 77, and (e) 83?

NEUTRON

The nucleus of an atom contains sub-atomic particles called *nucleons*. *Nucleons* are divided into two kinds of particles, *neutrons* and *protons*. Protons make up the dense nucleus core, but when chemists made calculations based on atomic weights of atoms, the numbers didn't add up. They knew there must be something they were missing. This is when *neutrons* were discovered. Neutrons are nuclear particles that have no charge and are located inside the crowded nucleus with positively charged protons.

> **Neutrons** are sub-atomic particles with a similar mass to their partner proton in the nucleus but with no electrical (+ or –) charge.

Table 5.1 shows common characteristics of electrons, protons, and neutrons.

Table 5.1 Common characteristics of electrons, protons, and neutrons indicate their special nature.

Name	Symbol	Mass (g)
Electron	e^-	9.110×10^{-28}
Proton	p^+	1.675×10^{-24}
Neutron	n	1.675×10^{-24}

Atomic Structure

Though Thomson, Rutherford, Meyer and Mendeleyev didn't quite understand what caused many of the reactions they observed, they recorded the patterns they saw among the elements. They noted that elemental properties seemed to reflect atomic weight and atomic number, but weren't really sure why. Modern chemistry has discovered the answers to these puzzles.

Figure 5.3 illustrates a beryllium atom with its energy levels. The atom is composed of 4 protons and 5 neutrons in the nucleus, and 4 electrons arranged in 2 shells (or orbital layers) outside the nucleus. The first shell contains 2 electrons and the second shell contains 2 electrons.

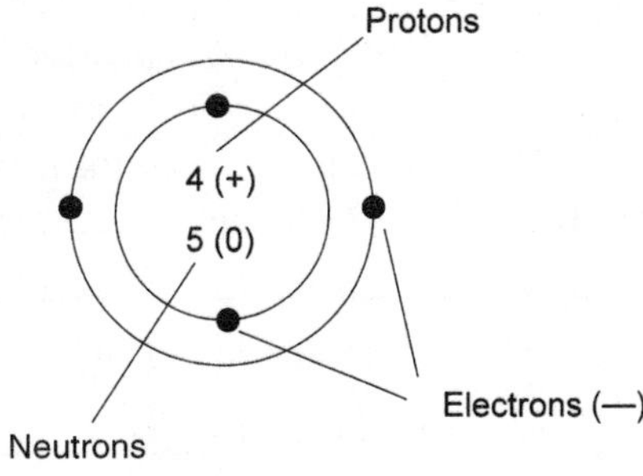

Fig. 5.3. Beryllium atom with its energy levels.

Through detailed experiments, scientists discovered that the way an element behaves is largely due to the number of electrons in its outer orbital shells. For example, elements with one electron in the outer shell behave alike; those with two electrons behave alike, and so on. This knowledge allowed early chemists to place similar elements in the same Periodic Table groups according to their outermost or *valence* electrons. We will learn more about electron bonding in the next chapter.

Molecules

Though many forms of matter like wood, rock, or soap appear solid upon first inspection, most matter is composed of a combination of atoms in a specific geometrical arrangement. The force that binds two or more atoms together is known as a *chemical bond.* A molecule is the basic joining of two or more atoms held together by chemical bonds. In a *covalent bond,* electrons are shared equally and in an *ionic bond,* electrons are transferred. We will see how this happens in Chapter 6.

Several elements occur naturally as two-atom or *diatomic* molecules. Of these, oxygen, nitrogen, hydrogen, fluorine, chlorine, bromine, and iodine occur in pairs at room temperature. These are in groups IA, VA, VIA, and VIIA of the Periodic Table. Such groupings make it easier for a researcher to tell a lot of specifics about an element at a glance. For example, those elements with one electron short of the noble gas structure are generally very electronegative. This means they really want to find or share an electron to form compounds by gaining an electron and completing a stable outer orbital.

Other molecules such as elemental phosphorus (P) are composed of four atoms, while sulfur (S) has eight atoms. Knowing the normal state of elements becomes important in predicting the outcome of chemical reactions.

If all the atoms within a molecule are composed of the same atoms, then it is a pure element. However, different elements combine with each other all the time. When this happens, then the product is called a *molecular compound.*

> A **molecule** is composed of the same kinds of atoms, chemically bonded by attractive forces.

One familiar compound is composed of two atoms of hydrogen and one atom of oxygen. This compound, water, written as H_2O is held together by covalent bonds. Compounds of combined element symbols are called *formulas* of the compound.

The formula for water is H_2O. The number of atoms of each element is written as a subscript in the formula, such as the 2 in H_2O. Later, when we take a closer look at the way atoms and elements combine, the importance of listing these subscript numbers will be more easily seen. (When no subscript is written, it is understood that only one atom of the element is involved.) The following example shows a few other molecular ratios in different compounds.

EXAMPLE 5.2

Some simple examples of the formulas of compounds are given.

Sodium chloride ($NaCl$) = 1 atom of sodium and 1 atom of chlorine.

Nitrous oxide $(NO_2)_2$ = 2 atoms of nitrogen and 4 atoms of oxygen.

Sucrose ($C_6H_{12}O_6$) = 6 atoms of carbon, 12 atoms of hydrogen, and 6 atoms of oxygen.

Since the earth contains many different forms of matter, solid, liquid, and gas, it is easy to see that atoms can combine in nearly infinite ways to form molecular compounds. However, there are only a certain number of discovered elements and sometimes chemical formulas are the same for different compounds. The way chemists keep these formulas straight is through their molecular and structural formulas.

Molecular Formula

A molecular formula is more specific than a compound's name. It gives the exact number of different atoms of each element in a molecule. We saw this earlier in the formula of water, H_2O.

Think of it as a closer look, like being shown the difference between a long-bed truck and an 18-wheel truck/trailer combination. The components are basically the same, engine, tires, body, and frame, but the number of wheels and length can make all the difference in the size and function of the vehicle.

In **Table 5.2** you will see some common molecular formulas.

A simple formula such as $CuSO_4$ gives the number of atoms of the different elements in the sample.

A molecular formula is more specific. It gives the exact number of different atoms of each element in a molecule. Water is written as H_2O, saltpeter (used

Table 5.2 Molecular formulas show the number of atoms present in a molecule.

Name	Chemical formula
Ethane	C_2H_6
Carbon tetrachloride	CCl_4
Oxalic acid	$H_2C_2O_4$
Cupric nitrate	$Cu(NO_3)_2$
Diphosphorus trioxide	P_2O_3
Ammonium nitrate	NH_4NO_3
Urea	NH_2CONH_2
Sulfuric acid	H_2SO_4
Calcium hydroxide	$Ca(OH)_2$
Sodium stearate	$C_{18}H_{35}O_2Na$
Benzene	C_6H_6

in fireworks and fertilizer) is KNO_3, and fructose (the sugar found in fruit and honey) is $C_6H_{12}O_6$.

An element is composed of the same kinds of atoms, chemically bonded by attractive forces. These atoms are usually held together in a certain way. Within a specific element, its atoms combine in certain ways with its own atoms, as well as those of other elements. This bonding comes about because of the properties of electrons and their location around each atomic nucleus.

Structural formulas show how specific atoms are ordered and arranged in compounds.

Think of it like a football game. The plays are set up with different players placed in certain positions. Each play is designed to serve a particular purpose. If the players form up one way, the quarterback may throw the ball. Set up in another way and the end player runs the ball over and across. If the players on the other side don't react to a certain configuration in the predicted way, the quarterback may have to run the ball. Placement and function of individual players is everything in football.

The same is true of chemistry. The arrangement of the atoms in a molecule can make a big difference in the characteristics and reactivity of compounds.

Figure 5.4 shows structural formulas with individual elements indicated. A structural formula shows exactly how an element is connected to the others in the molecule. Researchers study the structure of a molecule to figure out how it will react in a reaction. Structure has a definite effect on the properties of compounds. Knowing compounds' structures and their bonding abilities will make it easier to understand how molecules do what they do. We will see more of the kinds of bonds that can be formed in later chapters.

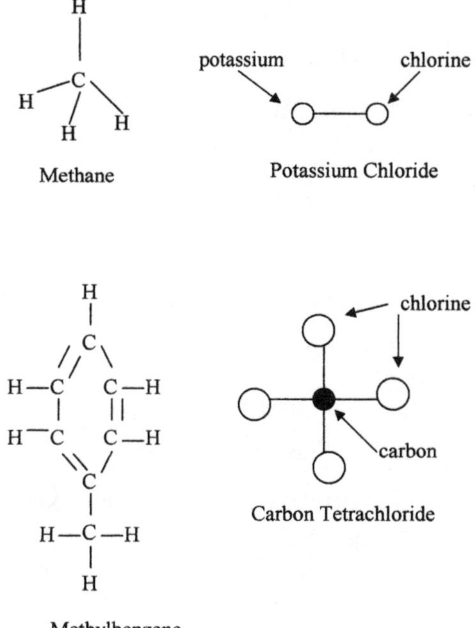

Fig. 5.4. Structural formulas of different compounds make it easier to see how atoms are bonded.

Quiz 5

1. Electrons are
   (a) sub-atomic particles with a +1 charge
   (b) equal to the number of protons in a nucleus
   (c) said to be charmed
   (d) sub-atomic particles with a +2 charge

2. The "plum pudding" model of an atom was replaced by the
   (a) orbital theory of protons and neutrons
   (b) "plum pudding" model with currants
   (c) Meyer model of electrons
   (d) Octave rule

3. The modern concept of the atom was developed by
   (a) E. W. Muller
   (b) J. J. Thomson
   (c) Lothar Meyer
   (d) Ernest Rutherford

4. Protons share the dense nucleus with
   (a) neutrons
   (b) molecules
   (c) quarks
   (d) atoms

5. The arrangement of atoms in a molecule can
   (a) cause explosive reactions
   (b) affect the reactivity of a molecule
   (c) mean the difference between hadrons
   (d) predict when it will be discovered

6. The force that binds two or more atoms together is known as
   (a) a chemical bond
   (b) valence
   (c) joules
   (d) electromagnetism

7. A structural formula shows
   (a) the calculated distance between atoms
   (b) the molecular weight of a molecule
   (c) the valency of oxygen
   (d) how an element is connected to others in the molecule

8. An atom
   (a) is 10^{-8} Angstroms in diameter
   (b) is composed of different sub-particles
   (c) has only two electrons in each orbital shell
   (d) is classified as strange and charmed

9. A molecular formula
   (a) gives the total number of moles in a compound
   (b) is used only in derived chemical formulas
   (c) gives the number of each elemental atom in a molecule
   (d) is handy to have, but doesn't include all the elements

10. The neutron is a sub-atomic particle
    (a) with no electric charge
    (b) with a -1 electric charge
    (c) smaller than an electron
    (d) with no counterpart within the atom

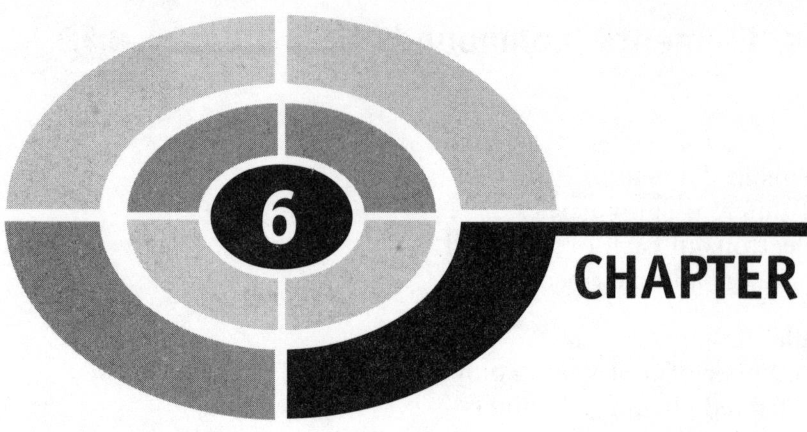

Electron Configurations

Electrons play a big role in the way the entire atom gets along with its friends, family, and neighbors. They don't stick together on a whim. Usually, atoms of an element will only combine with identical atoms or those of other elements in certain ways. Electrons serve as the glue between the nuclei of two or more atoms.

The word *electron* was coined by G. Johnstone Stoney in 1891. It was used to describe a unit of electrical charge measured in his experiments where an electrical current was sent through various chemical solutions to test its effect.

> A **chemical bond** is the attachment between atoms within a molecule.

The number of bonds an atom can form with other atoms depends upon the number of electrons it can easily share with its neighbors. The more electrons available to be shared, the more readily an atom will bond. In general, atoms combine in the numbers shown in **Table 6.1.**

Table 6.1 Atoms combine in certain numbers.

Element	Symbol	Number of possible bonds
Hydrogen	H	1
Carbon	C	4
Nitrogen	N	3
Oxygen	O	2
Chlorine	Cl	1

Elements can be placed in rows and columns of the Periodic Table by knowing something about their properties. The numbers at the top of the chart, Roman numerals I–VIII (or alternatively 1–18), are used to identify groups and chemical properties. The element groups usually give the number of electrons in the outermost orbital of the atoms in each column. Remember, these outer electrons are known as *valence* electrons. For example, the atoms of the elements in column IV have four electrons available to create bonds. Elements in column II have two free electrons in the outermost orbit around the nucleus.

Just as John Newlands overlapped elements around a cylinder and saw repeating patterns of characteristics, so too do elements in each family change in predictable ways. As you look down the columns of the Periodic Table, these characteristics can be compared. **Table 6.2** shows the pattern of a few elements from the alkali metal family.

Table 6.2 There are common properties in the alkali metal family.

Element	Melting point (°C)	Boiling point °C	Density (g/cm^3)
Li	180.5	1347	0.534
Na	97.8	881.4	0.968
K	63.2	765.5	0.856
Rb	39.0	688	1.532
Cs	28.5	705	1.90

Electron Configuration

The *electron configuration* of an atom describes the specific dispersal of electrons among available subshells.

The capacity of an energy level can be found in the following formula:

$$2n^s$$

where n is called the *principal quantum number* and indicates the energy level. In the first energy level, $n = 1$; the second energy level $n = 2$; the third energy level $n = 3$; and so on. **Figure 6.1** shows the electrons in these different orbital subshell energy levels.

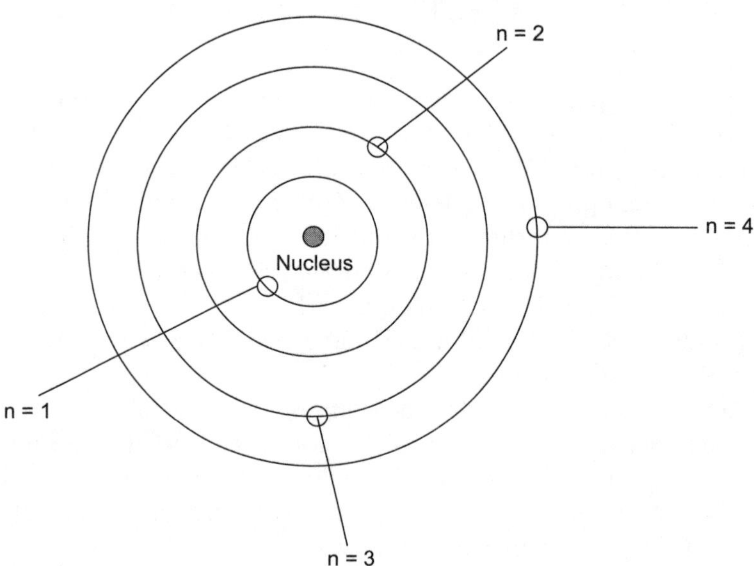

Fig. 6.1. Orbital subshell energy levels allow the chemist to figure out bonding of elements.

Electron subshells or orbitals are written with s, p, d, and f as terms for each known energy level. If you were to write the electron configuration of calcium (Ca), with an atomic number of 20, it would look like the following: $1s^2\ 2s^2p^6\ 3s^2p^6\ 4s^2$. It has 2 electrons in the 1s subshell, 8 electrons in the 2sp, 8 electrons in the 3sp, and 2 electrons in the 4s subshell. Another way of writing this is that there are 2 electrons in the $n = 1$ level, 8 electrons in the $n = 2$ level, 8 electrons in the $n = 3$ level, and 2 electrons in the $n = 4$ level.

The overall capacity of the energy levels then looks like that shown in **Table 6.3**.

Table 6.3 Electron energy levels have specific capacities.

Quantum number designation (n)	**Shell capacity ($2n^2$)**
1	$2 \times 1^2 = 2$
2	$2 \times 2^2 = 8$
3	$2 \times 3^2 = 18$
4	$2 \times 4^2 = 32$

Orbitals of the s-type are always singular, p-types form orbital sets of 3, d-type orbitals come in sets of 5, and f-type orbitals are written in sets of 7.

The 4s electrons of calcium are found in the outermost orbit to be filled and from this position react with other elements. Valence electrons affect the reactivity of atoms with other elements.

Molecules that share electrons are generally smaller, have lower melting and boiling points, are insoluble in water, and do not conduct electricity. The s and p orbitals of nearby atoms overlap to form a mixed orbital.

EXAMPLE 6.1

A simple example is that of oxygen (O). The atomic number of oxygen is 8. The electron configuration of oxygen is:

$$1s^2 2s^2 p^4$$

2 electrons in the 1s subshell, 2 electrons in the 2s subshell, and 4 electrons in the p subshell give a total of 8.

An *orbital diagram* (shown as circles) is the notation used to show the number of electrons in each subshell. Each subshell is labeled with its subshell notation, s, p, d, or f. An orbital diagram also makes it easy to see the sequence of how subshells are filled. If you use small circles to stand for a subshell, then the orbital diagram can be used to find the orbital configuration of nearly every element. **Figure 6.2** shows the order of an orbital filling sequence.

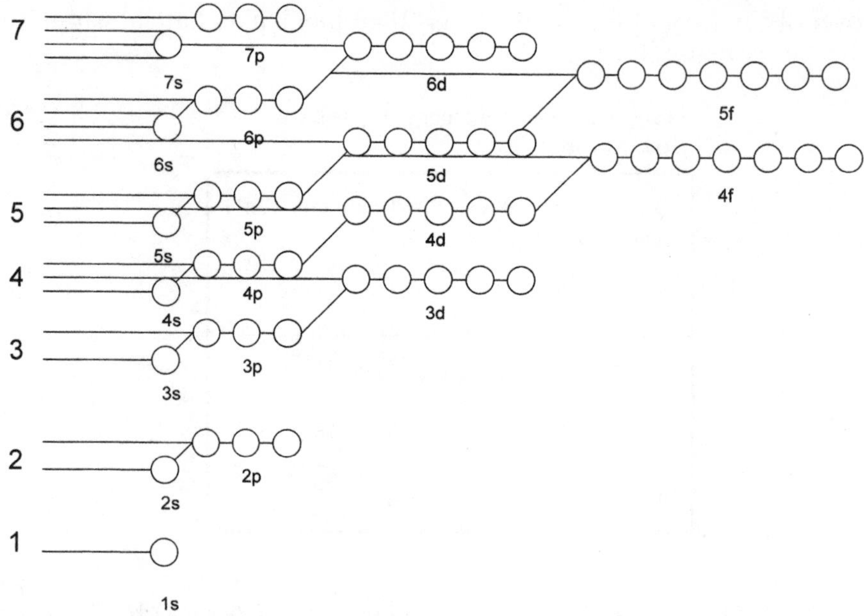

Fig. 6.2. Orbitals fill in a regular sequenced order.

Ground State

Atoms contain an infinite number of possible electron configurations. The configuration associated with the lowest energy level of the atoms is called the *ground state*. When an atom's energy levels go from a lower energy to a higher energy level, the change is sometimes seen as a flash of color or sudden heat.

> The **Aufbau principle**, also called the building-up principle, is used to show an atom's ground state.

Atoms fill up subshells like eggs in a carton. An atom's ground state adds electrons in a particular building order.

The standard building order for the description of an atom's orbital configuration is 1s, 2s, 2p, 3s, 3p, 4s, 3d, 4p, 5s, 4d, 5p, 6s, 4f, 5d, 6p, 7s, and 5f.

An example of an atom's orbital configuration is illustrated in **Figure 6.3** for the element magnesium.

Mg

Magnesium

Atomic Number – 12

Atomic Mass – 24.31

Group – II

Period – 3

Metal

Electrons per orbital layer – 2,8,2

Valence electrons – $1s^2\ 2s^2p^6\ 3s^2$

Fig. 6.3. The orbital configuration of magnesium shows how it can combine with other elements.

The Aufbau principle shows how energy increases in subshells. By filling the lowest energy subshells first, the ground state is built up. As the energy of the atom increases, the number of subshells filled increases.

When the number of electrons in an atom equals the atomic number (the number of protons), then the atom is neutral. Electron configurations are seen in the Periodic Table under the atomic number and element symbol.

The noble gases are in the ground state with the $1s^2$ (He) and p orbitals filled (Ne, Ar, and Kr) so they usually have no interest in reacting with anything else. In **Figure 6.4** the Periodic Table is shown with the location of the element orbitals.

1s													1s
2s										2p			
3s										3p			
4s					3d					4p			
5s					4d					5p			
6s					5d					6p			
7s					6d								

					4f						
					5f						

Fig. 6.4. Periodic Table with electron orbitals.

Orbital Filling

While some scientists were studying the energy levels as element energy increased, Friedrich Hund worked on figuring out the lowest energy that electrons could be arranged in a subshell.

> **Hund's rule** states that all orbitals of a given sublevel must be occupied by a single electron before pairing begins.

EXAMPLE 6.2

A simple example is carbon, atomic number (Z) = 6, orbital configuration of $1s^2 2s^2 2p^2$:

EXAMPLE 6.3

A more complex example is vanadium, a metal additive to steel, with an atomic number (Z) = 23, orbital configuration of $1s^2$ $2s^2p^6$ $3s^2p^6d^3$ $4s^2$:

Quantum State of Electrons

The Austrian physicist, Wolfgang Pauli, won the Nobel Prize for Physics for his exclusion principle; that no two electrons can have the same quantum state (position, momentum, mass, and spin) simultaneously.

> The **Pauli exclusion principle** states that no two electrons in the same atom can be in the same configuration at the same time.

Since each orbital can only have two electrons and those must be opposite in charge (or spin), then there are only two possible values for m_s. Look at the examples on the next page. Are all the orbital diagrams possible?

EXAMPLE 6.4

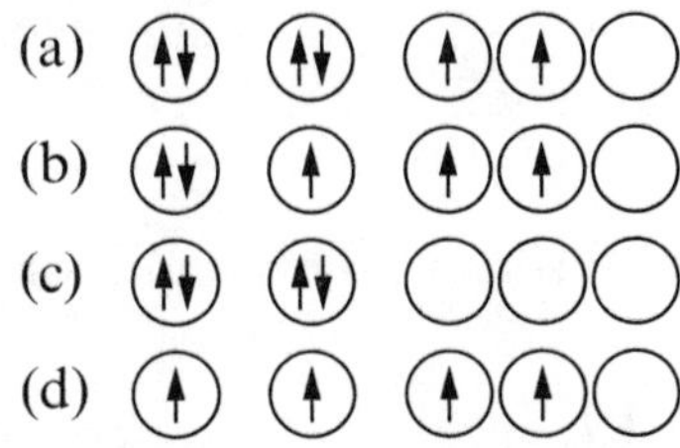

Did you get (a) yes, (b) no, (c) yes, (d) no, and (e) no?

Spin Magnetism

In 1921, two scientists, Otto Stern and Walter Gerlach, experimenting with silver atoms and a specially designed magnet found that electrons act like tiny magnets themselves. **Figure 6.5** shows how. By sending a beam of atoms through a magnet, the beam is split into two beams, one bending one way

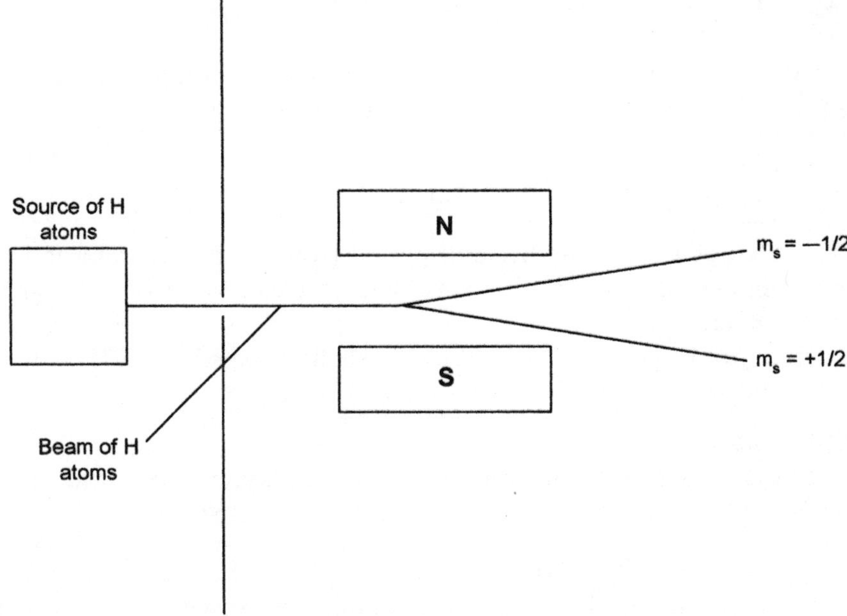

Fig. 6.5. Electrons attracted and repelled like magnets.

and the other bending the other way. They decided that the electrons must be attracted and repelled by opposite and like charges just like magnets attract and repel opposite and like charges. This was called *spin magnetism.*

EXAMPLE 6.5
This example shows an orbital diagram of electrons in the s and p orbitals.

The electrons in orbital diagrams are written as up and down arrows for $\uparrow m_s = +1/2$ and $\downarrow m_s = -1/2$. Electrons spinning through their orbitals act like spinning marbles of electrical charge. This spinning electrical charge circulates in the orbitals creating mini magnetic fields. This *spin magnetism* value is written as m_s.

Additionally, just as opposites charges attract and like charges repel, so it is with pairs of electrons in orbital subshells. Two electrons with the same spin cannot be placed in orbitals together. If one electron is spinning positively, then the other must be spinning negatively.

Subshells and the Periodic Table

The Periodic Table helps chemists write the atomic makeup of a molecule fairly simply. The arrangement of a main group element uses the following formula:

$$ns^a np^b$$

where $n =$ the outer shell number and period (row) of the element and $a + b =$ total number of valence electrons that you can get from the group (column) number.

In general, groups IA and IIA fill the s subshell, groups IIIA–VIIIA fill the p subshell, and groups IIIB–VIIIB fill the $(n - 1)$ d subshell.

EXAMPLE 6.6
Using the period and group information of the Periodic Table, what is the configuration of phosphorus (P), atomic number $Z = 15$?

Start with the first subshell on the Periodic Table as 1s, then in the second period (row) you have 2s. Jumping across in the same row is 2p. In the third period, there is 3s, 3p. In the fourth period there is 4s, 3d, and 4p.

$1s^2$ (first period) $2s^22p^6$ (second period) $3s^23p^3$ (third period)
Phosphorus is in period 3 so $n = 3$ and group 5A so valence electrons = 5 (Remember, the valence electron shell arrangement is the same as the outermost placement of electrons.)

EXAMPLE 6.7
Try nickel (Ni), atomic number $Z = 28$. First consult the Periodic Table for the period number and group. Then start building the subshells. Did you get $1s^22s^22p^63s^23p^64s^23d^8$? This can also be written with the subshells grouped together as $1s^2\ 2s^2p^6\ 3s^2p^6d^8\ 4s^2$.

EXAMPLE 6.8
If cesium (Ce) has a configuration of $1s^2\ 2s^2p^6\ 3s^2p^6d^{10}\ 4s^2p^6d^{10}\ 5s^2p^6\ 6s^1$ what is its period number and group? What is the filling valence subshell?

Cesium is an alkali metal in period 6 and group 1. The filling orbital is $6s^1$.

It is important to remember that electrons of different elements in the same groups look the same with the exception of different orbital configurations. Members of the same family have identical valence electron structures, when considering numbers of paired and unpaired valence electrons. These similar valence structures make it possible for family group members to react similarly.

Additionally, the elements with the most orbitals, further and further out from the nucleus, become increasingly more reactive as the electrons zip through a larger area. They have more "party room" to come in contact with other atoms and they take advantage of it!

Ionization Energy

The period of an element also describes patterns. If you look at the elements from left to right in the Periodic Table along any row

> The **ionization energy** of an element is the energy needed to detach an electron from an atom of an element.

Regular property changes can be compared to changes in electron arrangement. The higher the number of electrons in the outermost shell of an atom, the higher the ionization energy of that atom.

The elements are listed in rows so that it is easy to find information quickly. For example, if you were interested in a specific element, you would check out its place in the Periodic Table. Who are its neighbors on the chart? Which group is it in? Which period? How many electrons are in its outermost orbit (sometimes called outermost shell)? Is it reactive or not? Is it a metal or non-metal? Look again at Figure 6.4. All the group and period information that you will need on the elements can be found on the Periodic Table.

Knowing the reactivity of an element is important. If the element to be studied was potassium and you put it into water, you would have a wild reaction since alkali metals get really crazy in water. They give off hydrogen gas that ignites with the heat of the reaction and gives off a violet flame, like a lot of mini-fireworks. From potassium's soft solid state and its low boiling point, you might think it is a mild-mannered element. However, the reactive heat from an encounter with water changes a solid chunk of potassium into a liquid by melting it. Can't judge an element by its cover!

We will see more of the bonding preferences and special qualities of different elements in Chapter 13, when we talk about chemical bonding.

Quiz 6

1. The Aufbau principle
   (a) defines the undefined particles of the nucleus
   (b) is a method used to describe an atom's ground state
   (c) provides radioactive levels of elements
   (d) lists the negative and positive spin of atoms

2. Electrons have
   (a) a positive charge
   (b) are unreactive in the metal group
   (c) serve as the glue between nuclei of atoms
   (d) have only two outermost orbits

3. Ionization energy of an element
   (a) is the amount of thrust needed to fly at mach speeds
   (b) is the energy a neutron generates
   (c) cannot be calculated or observed
   (d) is the energy needed to detach an electron from an elemental atom

4. Noble gases
   (a) are highly reactive with helium
   (b) are highly reactive with strontium

(c) are unreactive under normal conditions
(d) are not related to nobility

5. A bond between atoms in a molecule is
(a) made up of a shared electron pair
(b) stronger than static electricity
(c) always located in the 3s orbital
(d) only a double bond

6. Elements in column IV of the Periodic Table have
(a) three electrons with which to create bonds
(b) four electrons with which to create bonds
(c) five electrons with which to create bonds
(d) unreactive bonding electrons for other elements

7. The electron configuration of an atom
(a) is determined by the amount of kinetic energy present
(b) is found by calculating atomic mass
(c) is written as s, p, d, and f subshells
(d) describes the specific distribution of electrons in a subshell

8. Friedrich Hund worked on
(a) calculating the energy signature of calcium
(b) the nature of electron spin
(c) the lowest energy arrangements of subshell electrons
(d) his family's tulip farm until he was twelve

9. The number of bonds an atom can form with other atoms
(a) depends on its overall size
(b) is calculated using orbital theory
(c) depends on the specific gravity of the atom
(d) depends on the number of electrons it can share

10. The Pauli exclusion principle states that
(a) any atoms with a free s orbital can form bonds
(b) no two atoms can occupy the same orbital unless their spins are different
(c) two atoms sharing an orbital are matched exactly
(d) atoms of the same configuration do not change

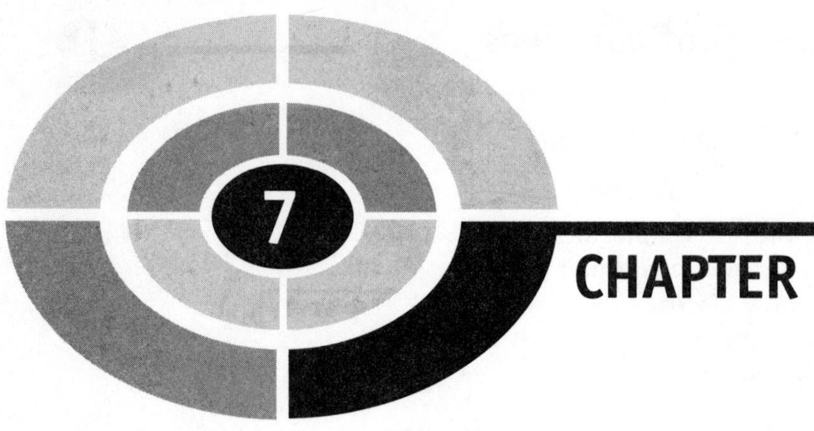

Concentration and Molarity

In the laboratory, chemists must make use of what they have on hand. It would be a waste of time to run out and buy 20 different concentrations of hydrochloric acid, for example, when they could dilute down the concentrated hydrochloric acid on the shelf and get whatever lower concentration was needed.

Solutions

A *solution* is a homogeneous mixture of a *solute*, the element or compound that dissolves, in a *solvent*, the solution that disperses the solute. Solutes and solvents can be elements or compounds. Commonly, the larger sample is called the solvent and the smaller sample is called the solute. When salt is dissolved in water, the salt is the solute and water (known as the *universal solvent*) is the solvent.

A **solute** is an element or compound dissolved to form a solution. A **solvent** is a liquid to which an element or compound has been added to form a solution.

To understand how atoms, elements, and molecules interact in solutions, it is important to understand concentration. You have seen how a high concentration of an element or solution can be much different from a low concentration. Consider salt water. Different concentrations can have very different effects. A physiological saline, 0.85% NaCl solution, given intravenously and used to stabilize patients' fluid levels in hospitals, is much different from the highly concentrated salt brine (30–50% NaCl) used to preserve codfish. In fact, confusing the two could be deadly.

When two or more liquids are able to form a solution, they are called *miscible*. Alcohol is miscible in water. In fact, there is an entire area of study in physics called fluid dynamics that tests the miscibility of fluids and the way they flow at different concentrations.

When two liquids don't mix, like oil and water, they are *immiscible*. Picture a lava lamp. There are generally large round globules of "lava," made of a specially compounded and colored wax, floating lazily around in a specially formulated liquid (most likely water). This mixture is heated with a light bulb in the base of the lamp that causes the lava to heat and expand until it becomes less dense than the contained liquid above. When this happens, the lava begins to rise to the top of the container where it slowly cools. When the lava is cool enough to sink to the bottom, it is heated again and the process is repeated. This is truly an example of chemistry as entertainment.

A *colloid* is like a homogeneous solution, but it is made up of larger particles of one solution mixed and spread all through another solution; think of chocolate chip cookie dough. There are two parts to a colloidal mixture, the *dispersing medium* and the *dispersed phase*. The dispersing medium is the substance in a colloidal mixture that is in the greater amount, like the cookie dough. The dispersed phase is the substance in the colloidal mixture that is in the smaller amount, like the chocolate chips.

Colloidal mixtures are not true solutions. Remember, in solutions the solute dissolves in the solvent. In colloidal solutions, the components don't dissolve, they just mix. When the parts of a compound mix in this way, they are said to have *colloidal* properties.

Percent (%) By Mass

It is often important to figure out how much of a mixture or *solute* has been added to the original solution or *solvent*. You can calculate this simply by finding the percent by mass of the solute.

To find the % of a specific compound in a solution, the formula is

$$\% \text{ by mass (mixing in solute)} = \frac{\text{mass of solute you are mixing}}{\text{mass of total solution}} \times 100\%$$

EXAMPLE 7.1

Using the saltwater example above, let's figure out the % by mass of sodium chloride (NaCl), if 2.35 g of NaCl is dissolved in 7.45 g of water. First, what is the total mass of the solution? What is the % of NaCl? The total mass of 2.35 g NaCl (solute) and 7.45 g water (solvent) = 9.80 g solution. Then if you divide the 2.35 g NaCl solute by the 9.80 g solvent and multiply by 100%, you will get 23.9795 or 24% NaCl by mass.

Parts Per Million

Sometimes we hear of a chemical compound mistakenly released into the air. To decide whether or not to evacuate the surrounding area, scientists and officials first need to find out the type of compound released and its properties. For example, a gas may be fine when contained, but upon release interacts with oxygen in the air and produces a nasty brew.

In and around cities, there is the extra problem of industrial releases combining with already present environment pollutants (think smog) such as carbon dioxide (CO_2), sulfur dioxide (SO_2), sulfuric acid (H_2SO_4), nitrogen oxide (NO_2), and nitric acid (HNO_3).

After finding out what the release is made of, the chemical dose that someone passing by on the street might be exposed to must be determined. The higher the dose, the greater the effect and sometimes the greater the risk to life and limb. Though many industrial releases are minimal, some chemicals are highly toxic at even the very lowest levels.

Most environmental releases are measured in the parts per million (ppm) or parts per billion (ppb) range. Such low levels make it hard to know if there is a problem. Chemicals released into the air or water may show up weeks or years later to be dangerous even when diluted.

A pinch of chilli pepper in 10 tons of beans would roughly be the equivalent of one part per billion. Such small levels seem insignificant and practically undetectable, but a chemical's reactivity must also be considered in order to know for sure if a release of a hundred parts per billion presents a problem.

EXAMPLE 7.2

Parts per million can be found by multiplying the ratio of the mass of solute to mass of solution by 10^6 ppm instead of 100%.

$$\frac{4.2 \times 10^{-3}\ \text{g (solute)}}{1.0 \times 10^{3}\ \text{g (solution)}} \times 100\% = 4.2 \times 10^{-4}\%$$

$$\frac{4.2 \times 10^{-3}\ \text{g}}{1.0 \times 10^{3}\ \text{g}} \times 10^{6}\ \text{ppm} = 4.2\ \text{ppm}$$

$$\frac{4.2 \times 10^{-3}\ \text{g}}{1.0 \times 10^{3}\ \text{g}} \times 10^{9}\ \text{ppb} = 4.2 \times 10^{3}\ \text{ppb}$$

In the above example, the chemical release would probably be reported in ppm.

Table 7.1 lists some ppm levels of common chemical contaminants.

Table 7.1 Pollutants are most often measured in parts per million in the air and water.

Pollutant	**Toxic levels (ppm)**
Arsenic in playground soil	10.0
Arsenic in mine tailings (toxic)	1320
Diethyl ether	400
Trihalomethane (in water)	0.10
Nitrate (in water)	10.0
Nitrite (in water)	1.0
Mercury (in water)	0.002
Cadmium (in water)	0.005
Silver (in water)	0.05

Molarity

Many times the exact amount of a solute is required for a specific volume of solution. Researchers aren't able to repeat their own experiments, let alone someone else's, unless exact quantities and measurements are made. In order to improve accuracy, the concentration of a solution is given in *molarity*.

> **Molarity** (M) equals concentration. M equals the number of moles of solute (n) per volume in liters (V) of solution.

A *mole* in chemical terms is not a small furry animal that lives in dark underground burrows, but an SI unit amount of a sample. One mole is defined as being the amount of sample having as many atoms (or molecules or ions or electrons) as carbon atoms in 12 grams of carbon.

Avogadro's Number

In 1811, Italian physicist, Amedeo Avogadro, presented a theory in the *Journal de Physique* that the mass of one mole of a sample contains the same basic number of particles as in 12 grams of ^{12}C. That number of atoms is called *Avogadro's number* or a *mole* of sample.

> **Avogadro's number (N)** or one **mole (mol)** is equal to 6.02×10^{23} atoms or molecules.

EXAMPLE 7.3

If you wanted to find the number of molecules in 1.35 moles, you would multiply the number of moles by Avogadro's number.

$$6.022 \times 10^{23} \times 1.35 \text{ moles} = 8.1297 \times 10^{23} \text{ molecules}$$

Avogadro's Law

That same year, Avogadro proposed that equal volumes of different gases at the same pressure and temperature all contain the same number of particles. Like Avogadro's number, this observation was used to solve chemical bonding and molecular composition problems.

To better understand the size of Avogadro's number, think of the size of a pea. If one mole of peas were spread over the earth's surface, then the surface would be covered by a layer nearly 20 miles thick. A high-speed computer can count all the fish in the sea in less than a second, but would take over a million years to count a mole of fish.

Atomic Mass

The *atomic mass* for a given element in the Periodic Table is measured out in grams equal to one mole of atoms of that element. The *molar mass (MM)* of elements and compounds is the mass, in grams, equal to the atomic and formula masses of those elements and compounds. Molar mass is measured in grams/mole.

Conversion of Mass to Moles

In the course of an experiment, it is usually necessary to figure out the mass of an element. For example, you might be asked to calculate the mass of 2 moles of potassium (K). To do this, you need a conversion factor that changes moles to mass. The conversion of moles to mass would look like this:

$$2.0 \text{ mol K} \times 39.0 \text{ g/mol K} = 78.0 \text{ g}$$

How about finding the number of moles in 124 grams of calcium? Mass to moles then would be mol/g:

$$124 \text{ g} \times 1 \text{ mol Ca}/40.1 \text{ g} = 3.09 \text{ mol Ca}$$

Using these simple conversion factors, then, a chemist is able to find the *molar mass* of a compound. The following solution of $CaCO_3$ demonstrates how it works:

$$Ca = 1 \text{ mol Ca} \times 40.1 \text{ g/mol Ca} = 40.1 \text{ g}$$

$$C = 1 \text{ mol C} \times 12.0 \text{ g/mol C} = 12 \text{ g}$$

$$O = 3 \text{ mol O} \times 16 \text{ g/mol O} = 48.0 \text{ g}$$

$$\text{Molar mass} = 100.1 \text{ g/mol } CaCO_3$$

To see the % of each element in $CaCO_3$, you would take the mass of each element in the one mole sample and then divide by the molar mass.

$$\% \text{ Ca} = 40.1 \text{ g Ca}/100.1 \text{ g } CaCO_3 \times 100 = 40.1\%$$

$$\% \text{ C} = 12 \text{ g C}/100.1 \text{ g } CaCO_3 \times 100 = 12.0\%$$

$$\% \text{ O} = 48 \text{ g O}/100.1 \text{ g } CaCO_3 \times 100 = 48.0\%$$

Concentration

Sometimes an experiment requires a weaker acid solution than what a chemist has on the shelf. In order to do the experiment, the solution must be diluted. This is done by figuring out the molarity and volume of the solution.

n_d = the number of moles in the dilute solution

v_d = the volume of the solution

M_d = the molarity of the solution

To figure out the number of moles of solute in the dilute solution, use

$$M_d \times v_d = n_d$$

To figure out the number of moles of solute in the concentrated solution, use

$$M_c \times v_c = n_c$$

The trick is that the number of moles does not change.

$$\text{Moles of solute} = n_d = n_c$$

So the equation looks like $M_c \times v_c = M_d \times v_d$

$$M_d = M_c \times v_c/v_d$$

As you begin working in a chemistry lab, you will use this formula more. For now, just an understanding of the relationship between solute and solvent is enough.

Dilution

Solutions can be found in their concentrated forms in many laboratories. They have a variety of uses, but, in general, it is easier to keep one concentrated solution on hand than five dilutions of the same solution. There is just not that much space on the shelves. It is important to learn how to make a less concentrated solution from a concentrated one. Look at the following example.

EXAMPLE 7.4

What would you do if an experimental procedure called for 1 M of hydrochloric acid (HCl) and all you had in the lab was 12 M HCl? Could you use what you had on hand? Sure! Just prepare the 1 M HCl by measuring a volume 1/12 or 82 milliliters of the concentrated solution into 1 liter of distilled water. The final concentration is equal to 1 M HCl.

EXAMPLE 7.5

How about nitric acid (HNO_3)? What if you needed 1 M of nitric acid for an experiment and only had concentrated nitric acid (16 M) on hand? You would measure out 63 ml of the concentrated HNO_3, then add enough distilled water to equal 1 liter. The resulting solution would be equal to 1 M HNO_3. If you needed a 3 M solution, multiply the 63 ml by 3 to get 189 ml to add to water to bring it to 1 liter. The resulting solution would be a 3 M solution.

To make diluted solutions, chemists use volumetric flasks or beakers for accurate measuring. With a bit of practice, making dilutions of concentrated solutions right off the shelf will be a snap.

Some people compare laboratory chemistry with cooking in the kitchen. Sample preparation, dilution, and mixing are all done with measured care to produce a final product with specific characteristics and qualities. In other words, it might be a masterpiece or a mess, depending on how you follow the directions.

Quiz 7

1. Concentration
   (a) is important to understand chemistry
   (b) is the volume per velocity of molecular movement
   (c) can mean the difference between life and death
   (d) is found by mixing a weak acid and strong base

2. Percent mass of solution
   (a) is used to find the amount of solute in solvent
   (b) is used to weigh grams onto a scientific balance
   (c) never uses the total mass of the solute
   (d) can be achieved without knowing the atomic weight of the chemicals involved

3. Chemical concentrations are most often expressed as
   (a) ppc (parts per centimeter)
   (b) ppb (parts per billion)
   (c) pgs (parts per gram solvent)
   (d) ppm (parts per million)

4. Molarity
   (a) equals mass
   (b) is used to find general amounts
   (c) equals concentration
   (d) has the opposite function of polarity

5. A mole
   (a) has 10^2 atoms in a sample
   (b) has the same number of atoms in a sample as 12 grams of ^{12}C
   (c) is a small, black, furry rodent that lives underground
   (d) is seldom used in modern chemical calculations

6. Avogadro's number
   (a) is equal to 6.02×10^{23} atoms or molecules
   (b) is equal to 4.20×10^{23} atoms or molecules
   (c) is equal to 2.60×10^{23} atoms or molecules
   (d) is equal to 23.02×10^{6} atoms or molecules

7. Avogadro's number
   (a) was first suggested by Anatasia Avogadro
   (b) has never been proven decisively
   (c) is used to calculate the number of electrons in a sample
   (d) is equal to one mole

8. Empirical formulas
   (a) show the elements' proportions in a compound
   (b) indicate probable ideal gas combinations
   (c) are always the same as the molecular formula
   (d) give the whole number ratio of elements in a compound

9. Molar mass (MM) is measured in
   (a) moles/solute
   (b) grams/liter
   (c) grams/mole
   (d) moles/solvent

10. In order to convert mass to moles, you need
    (a) a calculator
    (b) a list of atomic masses
    (c) the boiling point of the solvent
    (d) a very sensitive scale

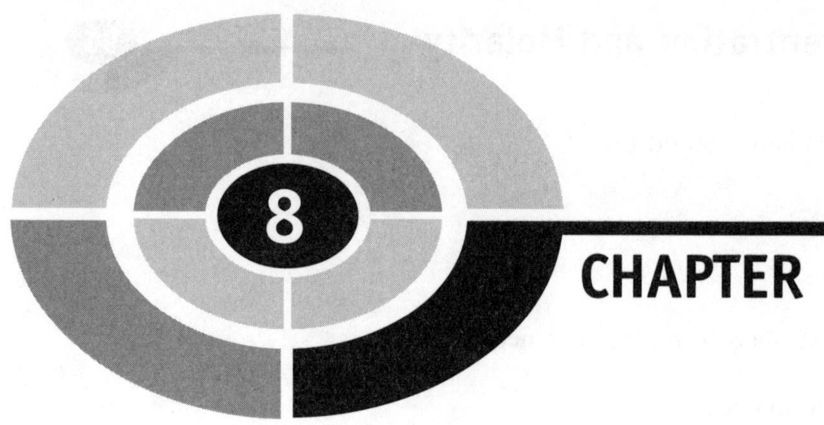

CHAPTER 8

The Hydrogen Atom

What is hydrogen anyway? We hear so much about this element. It may seem, at times, that it is a part of nearly every compound.

The hydrogen atom is a major player among the elements. It is easily the most abundant element in the universe and makes up about 90% of the universe by weight. Hydrogen is involved in most of the everyday compounds that we know of and is particularly important when bonding with carbon in organic chemistry.

Hydrogen gets its name from the Greek word *hudor*, which means water and *gennan,* which means to generate. Since the first chemical formula everyone seems to learn is H_2O, this is an easy one to remember.

Hydrogen was described in detail by Henry Cavendish in 1766. It is made up of one proton and one electron. With one proton, it has an atomic number of 1 and the honor of being the first element of the Periodic Table. A nonmetallic, colorless, tasteless, odorless gas at 298 K, hydrogen is highly flammable in the presence of oxygen.

Hydrogen got a bad reputation when engineers in 1937 used it as the lifting element for the huge airship Hindenberg. Later, research into the Hindenberg's crash and burn found that hydrogen was not the main culprit, but that static electricity had set fire to the aluminum-rich varnish of the airship's fabric covering, ignited the hydrogen within and caused the disaster.

Modern airships, sometimes called blimps for their oblong shape, use an unreactive gas like helium for lift. Helium is twice as dense as hydrogen, but still lighter than air.

Hydrogen, a chameleon among the elements, changes and reacts with most elements, especially carbon. Early investigators noticed its effects even before Cavendish wrote up the specifics of the element. They figured out that hydrogen and oxygen were separate elements of the compound water. Before this idea became accepted, everyone thought water, so limitless upon the earth, was a single element.

Besides being combined with oxygen in the form of water, hydrogen is also found in mines and oil and gas wells. Stars contain an almost limitless supply of hydrogen. Hydrogen is the most abundant element in the universe, making up over 90% of the visible universe's mass. Here on Earth, hydrogen is combined with oxygen to form water which covers over 70% of the Earth's surface. However, in the Earth's crust, hydrogen makes up only about 0.9% of the composition.

Hydrogen Compounds

Hydrogen changes and reacts with most elements, especially carbon, where it combines and forms starches, hydrocarbons, fats, oils, proteins, and enzymes. We will learn more about the hydrogen and carbon bonds in Chapter 10.

Hydrogen also reacts with nitrogen to form ammonia and the halogens to form acidic hydrogen halides. It combines with sulfur to form the rotten egg smell (hydrogen sulfide, H_2S), and with oxygen and sulfur to form sulfuric acid (H_2SO_4).

Other uses of hydrogen include the hydrogenation of oils, methanol production, rocket fuel, welding, the production of hydrochloric acid, and the reduction of metallic ores. Hydrogen is also important in cryogenics techniques and in superconductivity experiments since its melting point is only just above absolute zero.

Hydrogen Ion

The hydrogen ion is sometimes just called a "proton" in reactions since it carries a positive charge. In liquid form, the H^+ ion stays in the hydrated state (bonded with everything it can) such as the hydroxonium ion H_3O^+.

When Dalton was doing experiments to figure out the details of the atomic theory and whether or not one atom of an element had a particular mass, he burned hydrogen gas in oxygen to test his hypothesis. This experiment showed that 1 gram of hydrogen reacts with roughly 8 grams of oxygen. It wasn't until much later that chemists figured out that hydrogen actually bonded with 2 atoms of oxygen, atomic weight roughly 16 grams, to form the liquid we all know and love, water.

One of hydrogen's isotopes, tritium (^3H), is radioactive. Tritium is produced in nuclear reactors and is used in the production of the hydrogen bomb. It is also used as an additive agent in making shimmering paints and as a tracer isotope.

Reduction

Reduction is the chemical name for a decrease in oxidation number. When hydrogen is heated in combination with metal oxides like copper and zinc, the metal element is separated out and water is formed.

EXAMPLE 8.1
The example below demonstrates the reduction reaction.

$$CuO + H_2 \Rightarrow Cu + H_2O$$

$$ZnO + H_2 \Rightarrow Zn + H_2O$$

The metal oxide is *reduced* to release the uncombined metal. This is called a reduction reaction.

> **Reduction** is what happens when a sample gains one or more electrons provided by the reducing agent. As you might guess, hydrogen can be used as a reducing agent.

Reduction also happens when a compound picks up hydrogen atoms. Methyl alcohol, CH_3OH, is formed in the reaction of carbon monoxide (CO), hydrogen gas, and a catalyst.

EXAMPLE 8.2

$$CO + 2\ H_2 \Rightarrow CH_3OH$$

An atom is also reduced if it gains electrons directly. This can be seen when Cu^{2+}- or Ni^{2+}-containing solutions are plated onto an electrode. When electrons are removed from the electrode, the metal ions are reduced.

Oxidation

While studying reduction, it is a good time to look at the opposite reaction, *oxidation*. Oxidation is the process of a substance combining with oxygen. It is the reverse of reduction.

When silicon (Si) is combined with oxygen and other minerals in the Earth's crust, sand is formed. In fact, most of the compounds in living organisms on this planet contain oxygen. Nearly 60% of the weight of a human body is oxygen.

> **Oxidation** takes place when an element gains oxygen, loses hydrogen, or loses electrons.

EXAMPLE 8.3

$$N_2 + O_2 + \text{(high temperature)} \Rightarrow 2\ NO\ \text{(nitric oxide)}$$

or

$$2\ H_2S + 3O_2 \Rightarrow 2\ H_2O + 2SO_2$$

Hydrogen sulfide (H_2S) burns in oxygen to make water and sulfur dioxide. When this is oxidized, O_2 combines with each atom of the products. Oxidation and reduction can be thought of as an "opposite theory."

> ***Reduction* is (+) gain.**
> ***Oxidation* is (–) loss.**

The key to oxidation is to remember that oxidation is not just for oxygen, but it also defines a lot of different reactions with metals.

Figure 8.1 shows oxidation and reduction changes.

Basically, it is a cause and effect system. When an element causes the oxidation of another substance, it is itself reduced in the process.

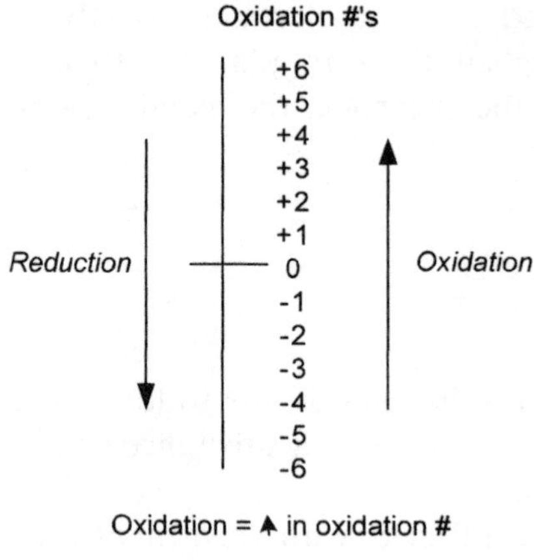

Fig. 8.1. Oxidation and reduction are opposite reactions.

> **Redox** chemistry is when one sample is reduced and another sample is oxidized at the same time during the same reaction.

Picture the "you scratch my back, I'll scratch yours," type of interaction. When all the atoms in a reaction shift around to find balance in filled orbitals, the compound is stable.

Oxidation and reduction are opposite reactions. An *oxidizing agent* is the element in the reaction that oxidizes another element, while at the same time being reduced itself. A *reducing agent* is the element in the reaction that reduces another element while at the same time being oxidized itself. In redox reactions, the total charge of the elements' oxidation numbers will be equal on both sides of the equation.

Half-reactions

In electrochemical reactions, you will sometimes see redox reactions written as half-reactions. The charge must be balanced overall and the electrons produced in one part of the reaction must be lost in another part. If this didn't happen, electrical charges would build up and cause a big problem.

Look at the electrical ion transfer (*electrolysis*) of molten sodium chloride to produce chlorine gas and sodium metal. The sodium (Na^+) and chloride

(Cl^-) ions are in the electrolyte solution. When an electrical current is passed through the solution, the chloride ions react at the anode and the sodium ions react at the cathode as shown in the following reactions:

$$2Cl^- \Rightarrow Cl_2 + 2e^- \text{ (oxidation)}$$

$$Na^+ + e^- \Rightarrow Na \text{ (reduction)}$$

The same number of electrons begins at the cathode and leaves at the anode, so the overall reaction is found by multiplying the second half-reaction by two and adding the two reactions to get:

$$2Na^+ + 2Cl^- \Rightarrow 2Na + Cl_2$$

Oxidation Number

The number used to keep track of the electrons in a reaction is called the *oxidation number*. Elements, like the halogens, may have different oxidation numbers depending on the specific reaction and environment. Oxidation number also tells us something about the strength or ability of a compound to be oxidized or reduced or to serve as an oxidizing agent or reducing agent. Oxygen has an oxidation number of −2. Using this as a starting place, oxidation numbers are assigned to all other elements.

Since water is a neutral molecule, it is fairly simple to figure out the oxidation numbers for the elements of the H_2O molecule. Oxygen has an oxidation number of −2, so then each hydrogen must have an oxidation number of +1, to allow the total charge to equal zero. **Table 8.1** lists a few hints to help figure out oxidation numbers.

Table 8.1 When figuring out oxidation numbers, remembering a few hints can help.

General rules of oxidation and reduction
1. An uncombined element has an oxidation number of zero (K, Fe, H_2, O_2) 2. All oxidation numbers added together in a compound must equal zero. 3. In an ion of one atom, the oxidation number is equal to the ion's charge. 4. In an ion of more than one atom, all the oxidation numbers add up to the ion's charge. 5. When oxygen is part of a compound, the oxidation number is −2 (except peroxides $H_2O_2(-1)$). 6. Hydrogen's oxidation number is equal to its +1 charge (except when combined with metals, then it is −1).

Oxidation numbers can be calculated by multiplying the number of elemental atoms present by the oxidation numbers and setting the entire equation equal to zero. The following example shows you how.

EXAMPLE 8.4

What is the oxidation number of chromium in $Li_2Cr_2O_7$, lithium dichromate?

$$Li = +1, Cr = x, O = -2$$

$$Li_2\ (2 \times +1) = 2$$

$$O_7\ (7 \times -2) = -14$$

$$+2 + 2x + (-14) = 0$$

$$2x = 12$$

$$x = 6$$

So the oxidation number of chromium is 6.

Figuring out the oxidation and reduction of elements in a sample is fairly simple if you use the Periodic Table and the rules of reaction. Working with redox reactions is basically an accounting task. You need to be able to keep track of all of the electrons as they transfer from one ion form to another. The trick to balancing redox reactions is to balance the charge as well as the elements on each side of the reaction.

In the next chapter, we will take a closer look at ions and how they act according to their specific characteristics.

Quiz 8

1. Hydrogen is key to
   (a) the making of candle wax
   (b) bonding with carbon in organic molecules
   (c) radioactive reactions
   (d) the formation of ozone

2. The atomic number of hydrogen is
   (a) 1
   (b) 2
   (c) 3
   (d) 4

3. In water molecules, how many oxygen atom(s) combine with hydrogen atoms?
   (a) 1
   (b) 2
   (c) 3
   (d) 4

4. Reduction is the process of
   (a) calculating the oxidation number of oxygen
   (b) increasing the oxidation number of hydrogen
   (c) losing hydrogen or electrons in a reaction
   (d) gaining hydrogen or electrons in a reaction

5. Hydrogen makes up roughly what % of the universe's visible mass?
   (a) 33%
   (b) 50%
   (c) 70%
   (d) 90%

6. Oxidation is
   (a) a (–) loss in oxidation number
   (b) a (+) gain in oxidation number
   (c) only possible with oxygen
   (d) a new form of acne medicine

7. When hydrogen sulfide burns in oxygen, the products
   (a) are oxygen and sulfur
   (b) are water and sulfur dioxide
   (c) smell like tea tree oil
   (d) smell like almonds

8. The number used to track electrons in a reaction is the
   (a) atomic number
   (b) reactant number
   (c) ionization number
   (d) oxidation number

9. What are the parts of "opposite theory"?
   (a) organic and inorganic elements
   (b) crystallization and condensation
   (c) oxidation and reduction
   (d) metals and non-metals

10. When one element causes the oxidation of another element, it is
    (a) oxidized
    (b) an acid
    (c) reduced
    (d) a base

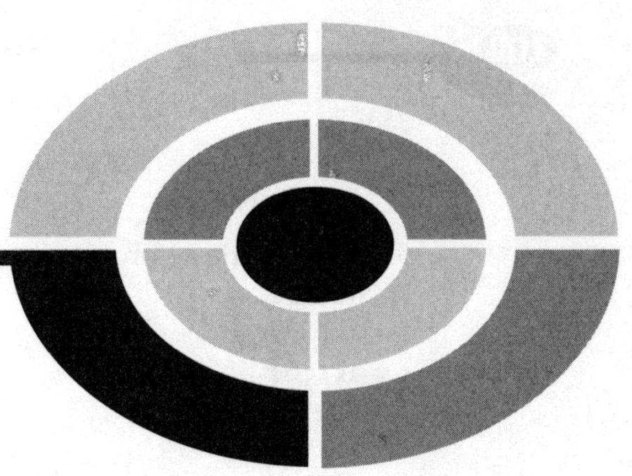

Test: Part Two

1. Who blasted particles through gold foil and found that atoms had positive centers?
(a) Thomas Edison
(b) Ernest Rutherford
(c) Steven Hawking
(d) Lothar Meyer

2. Nuclei are generally how big in diameter?
(a) 10^{-8} m
(b) 10^{-10} m
(c) 10^{-12} m
(d) 10^{-18} m

3. The molecular formula of saltpeter is
(a) CNO_3
(b) NaClO
(c) HNO_3
(d) KNO_3

4. Which of the following are not nucleons?
(a) protons
(b) dacrons

(c) electrons
(d) hadrons

5. Gold with 79 protons has an atomic number of
(a) 52
(b) 57
(c) 67
(d) 79

6. Hydrogen, fluorine, chlorine, bromine, and iodine are all what kind of molecules?
(a) diatomic
(b) monoatomic
(c) triatomic
(d) gases

7. Elemental phosphorus (P) is composed of how many atoms?
(a) 2
(b) 3
(c) 4
(d) 6

8. The position of elements in a molecule has the biggest impact on
(a) color
(b) reactivity
(c) length
(d) melting point

9. J.J. Thomson developed a model of the nucleus commonly called the
(a) tapioca model
(b) oatmeal raisin model
(c) flan model
(d) plum pudding model

10. Chemical formulas of molecules are
(a) always the same
(b) never the same
(c) sometimes the same
(d) not often used

11. If the temperature of a sample is 14°C, what is the temperature in kelvin?
(a) 224 kelvin
(b) 259 kelvin

(c) 287 kelvin
(d) 295 kelvin

12. A commonly used heat source in laboratory experiments is called a
(a) space heater
(b) Bunsen burner
(c) light bulb
(d) toaster

13. J.J. Thomson performed experiments in
(a) a cathode ray tube
(b) a venting hood
(c) a beaker
(d) a microwave oven

14. In Thomson's plum pudding model, the pudding is made of
(a) positively charged 'raisins'
(b) milk, butter, flour, and sugar
(c) a blob of positively charged particles
(d) quarks

15. What particle has a charge of 1.6×10^{-19} coulombs?
(a) xenon
(b) neuron
(c) proton
(d) electron

16. Orbitals of the p-type come in
(a) sets of 2
(b) sets of 3
(c) sets of 5
(d) sets of 7

17. The further out from the nucleus an electron orbital is located, the
(a) less reactive the electrons
(b) more orbital diagrams you have to write
(c) more reactive the electrons
(d) more problems they have finding their way home

18. The boiling point of lithium is
(a) 688°C
(b) 1347°C
(c) 1517°C
(d) 2212°C

19. Wolfram is the German name for which element?
(a) tin
(b) gold
(c) tungsten
(d) mercury

20. $1s^2\ 2s^2p^6\ 3s^2p^6\ 4s^2$ is the electron configuration for which element?
(a) calcium
(b) potassium
(c) nitrogen
(d) zirconium

21. Elements can be placed in rows and columns of the Periodic Table when you know something about their
(a) age
(b) date of discovery
(c) value
(d) properties

22. Electrons that affect the reactivity of atoms with other elements are called
(a) friendly electrons
(b) valence electrons
(c) strange electrons
(d) charmed electrons

23. If the shell capacity of an orbital is $2n^2 = 18$, what is n?
(a) $n = 1$
(b) $n = 2$
(c) $n = 3$
(d) $n = 4$

24. Electrons in orbital diagrams are written as
(a) horizontal squiggly lines
(b) thick black dots
(c) superscripts to the elements
(d) up and down arrows

25. When two or more liquids form a solution, they are
(a) always acids
(b) miscible
(c) immiscible
(d) liquids

26. Spin magnetism is when
    (a) electrons are attracted and repelled by opposite and like charges
    (b) news reporters report the news from a certain angle
    (c) disc jockeys play records well
    (d) elements change colors around magnets

27. A solution made up of larger particles of one solution mixed and spread all through another solution is called a
    (a) mess
    (b) science fair project
    (c) hydrophilic solution
    (d) colloidal solution

28. The number of bonds an atom can form with other atoms depends on
    (a) its color
    (b) the number of electrons it can share with its neighbors
    (c) the chemist
    (d) its melting point

29. In chemistry, water is
    (a) always a bad choice for washing hands
    (b) commonly called the universal solvent
    (c) written as HO
    (d) written as OH

30. In general, when making solutions, the solvent is
    (a) not typically used
    (b) smaller than the solute
    (c) larger than the solute
    (d) always equal to the solute

31. Hydrogen sulfide smells like
    (a) a locker room
    (b) vanilla
    (c) fermented fruit
    (d) rotten eggs

32. The "you scratch my back, I'll scratch yours" type of interaction describes
    (a) gorilla social skills
    (b) redox chemistry
    (c) alchemy
    (d) radioactivity

33. When an element causes the oxidation of another substance
(a) it is itself reduced in the process
(b) it is itself oxidized in the process
(c) there is never enough oxygen left to complete the reaction
(d) there is an over abundance of oxygen left to complete the reaction

34. The equation $CuO + H_2 \Rightarrow Cu + H_2O$, is an example of
(a) solidification
(b) vaporization
(c) oxidation
(d) reduction

35. Reduction describes
(a) electron (+) gain
(b) a helpful diet plan
(c) electron (–) loss
(d) a reduced spending plan

36. The density of cesium is
(a) 1.5 g/cm^3
(b) 1.9 g/cm^3
(c) 2.3 g/cm^3
(d) 2.9 g/cm^3

37. Oxygen is a common
(a) actinide
(b) ingredient found in anaerobic reactions
(c) oxidizing agent
(d) reducing agent

38. Stickstoff is the German name for which element?
(a) gold
(b) nitrogen
(c) aluminum
(d) arsenic

39. Hydrogen reacts with which element to form ammonia?
(a) oxygen
(b) carbon
(c) nitrogen
(d) sulfur

40. Hydrogen combines with carbon and forms all but one of the following?
    (a) hydrocarbons
    (b) proteins
    (c) starches
    (d) table salt

Elements, Groups, and Behavior

Atomic Number and Ions

To understand chemistry and chemical reactions, you have to understand the very basic building blocks of the elements. The protons, neutrons, and electrons are all pretty standard stuff by now. But there are more shortcuts to make the study of chemistry easier. One of them is atomic number.

Atomic Number

The modern Periodic Table contains over one hundred (still a matter of debate) elements. To better describe the central nature of the elements, the *atomic number* (Z) of the element is used. The atomic number (Z) is determined from the number of protons within the nucleus of an element. Neutrons are not recorded.

Atomic number (Z) gives us information about an element's character and allows us to balance the total electrical charge in a molecule.

> **Atomic number** (Z) is equal to the number of protons in the nucleus of an atom.

Remember when Mendeleyev's and Meyer's first attempts at arranging elements into the Periodic Table pointed to other undiscovered elements? At first, by atomic mass only, they arranged the elements on cards and shuffled them around in order to find the best fit. When elements were finally put into rows (periods) and columns (groups) by greater and greater atomic number, things got clearer.

Then, when they included the place holders they found similar functions and characteristics. Mendeleyev stuck with his intuition when placing cobalt and nickel, as well as tellurium and iodine, switching their places based on activity and character rather than mass only.

Gallium and germanium, discovered one year apart (1875 and 1876), slid into open slots as if Mendeleyev had been waiting for them to show up all along. Maybe he had.

Be careful not to confuse atomic number with atomic weight. Tungsten has 74 protons and its atomic number is 74. The atomic weight of tungsten is 183.85. Hydrogen has a single proton and its atomic number is 1. Lithium has 3 protons, 4 neutrons, and 3 electrons. Its atomic number is 3. Its atomic weight is 6.94. Mendelevium (named after you know who) has 101 protons and 157 neutrons, and 101 electrons. It has an atomic number of 101, but an atomic weight of 258. You get the idea.

EXAMPLE 9.1

See if you can figure out the following:

(a) What is the name of the element with the atomic number of 86?
(b) What is the name of the element with 37 protons?
(c) How many protons does yttrium have in its nucleus?
(d) What is the name of the element with three more protons than gold?
(e) What is the name of the element with one less proton than vanadium?

Did you get (a) radon, (b) rubidium, (c) 39, (d) lead, and (e) titanium?

When you find atomic number, you find the number of protons too! With this information, you can find a nucleus's charge. In general, atoms and molecules have equal numbers of electrons and protons and then have no overall charge. They are neutral.

All compounds are neutral and have zero charge. **Cations** (+) total overall charge = **anions** (–) total charge.

Ions

If an atom or molecule gains or loses one or more electrons, it is no longer neutral and becomes a charged *ion*. (*Note*: protons cannot change to become ions, only electrons.) Ions are charged elements with complete orbitals following the octet rule. They are unbalanced electrically, highly attracted to ions of the opposite charge, and eagerly stick together to form ionic compounds.

Research has found that most of the transition metals form more than one kind of ion with unique charges. Iron (Fe) forms two different ions, Fe^{2+} and Fe^{3+}, and nickel forms Ni^{2+} or Ni^{3+}. Cobalt (Co^{2+}, Co^{3+}), copper (Cu^{+}, Cu^{2+}), tin (Sn^{2+}, Sn^{4+}) and lead (Pb^{2+}, Pb^{4+}) also exist in more than one ion form.

MONATOMIC IONS

These ions are very simple, single ions that have a positive or negative charge. *Monatomic ions* can have fixed charges like hydrogen (H^{+}), cesium (Cs^{+}), sulfide (S^{-}), and nitride (N^{3-}) or variable charges like copper (Cu^{+}, Cu^{2+}), gold (Au^{+}, Au^{3+}), and tin (Sn^{2+}, Sn^{4+}). A good number of elements found in nature have had many different ion forms isolated.

POLYATOMIC IONS

When two or more atoms are covalently bonded, but keep an overall charge, they are known as *polyatomic ions*. Polyatomic ions are groups of atoms with an overall electrical charge. The majority of these ions are single element atoms bonded with oxygen. The element name is taken from the innermost or core atom. The single most common positive polyatomic ion is the ammonium ion (NH_4^{1+}). Positive polyatomic ions have a charge because the combined atoms from the two combining elements have lost one or more electrons.

Figure 9.1 shows a few monatomic and polyatomic cations and their charges.

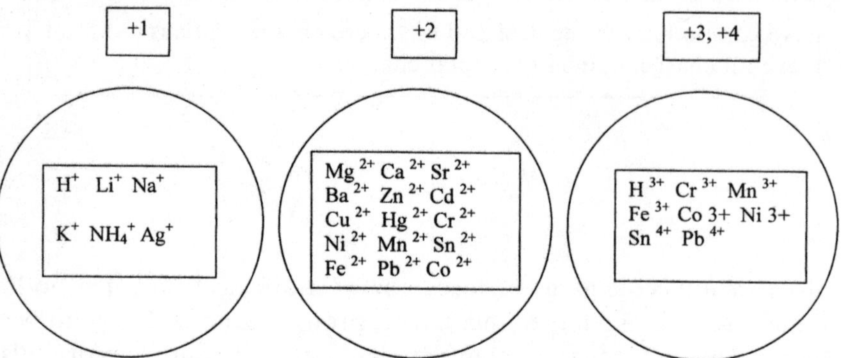

Fig. 9.1. Monatomic and polyatomic cations have specific charges.

Most of the common inorganic (no carbon) polyatomic ions have negative charges. The phosphate ion, PO_4^{3-}, has an overall negative charge as does the oxalate ion, $C_2O_4^{2-}$.

Table 9.1 lists some common monatomic and polyatomic anions (−).

> **Polyatomic ions** have two or more atoms held together by a covalent bond, but they still have an overall charge.

One way to recognize and study polyatomic ions and their specific elements is through their names. Chemical name endings are used with the idea of telling more about an ion's characteristics.

Ionic Compounds

Ionic *compounds* are held together by strong ionic bonds. For example, when a metal reacts with a non-metal, an ionic compound forms. Ionic compounds are commonly hard crystalline solids with high melting points due to the strength of their ionic bonds. Ionic compounds serve as *electrolytes* (they conduct electricity), when dissolved in water to form solutions.

NAMING IONIC COMPOUNDS

When naming an ionic compound, the first element gets the name root and the second adds on the ending "*ide*." The easiest example to remember is when sodium (Na) combines with chlorine (Cl) to form NaCl or sodium

Table 9.1 Monatomic and polyatomic anions are important to keep track of in reactions.

−1	−2	−3 and −4
Acetate $C_2H_3O_2^-$	Carbonate CO_3^{2-}	Arsenide As^{3-}
Acetate CH_3COO^-	Chromate CrO_4^{2-}	Borate BO^{3-}
Bromate BrO_3^-	Dichromate $Cr_2O_7^{2-}$	Carbide C^{4-}
Bromide Br^-	Hydrogen phosphate HPO_4^{2-}	Nitride N^{3-}
Chlorate ClO_3^-	Oxalate $C_2O_4^{2-}$	Phosphate PO_4^{3-}
Chloride Cl^-	Oxide O^{2-}	Phosphide P^{3-}
Cyanide CN^-	Selenide Se^{2-}	Phosphite PO_3^{3-}
Hydride H^-	Sulfate SO_4^{2-}	
Hydrogen carbonate HCO_3^-	Sulfide S^{2-}	
Hydrogen sulfate HSO_4^-	Telluride Te^{2-}	
Hydroxide OH^-	Thiosulfate $S_3O_3^{2-}$	
Iodide I^-		
Nitrate NO_3^-		
Perchlorate ClO_4^-		
Permanganate MnO_4^-		
Thiocyanate SCN^-		

chlor*ide*. Other ionic compounds include calcium ox*ide*, lithium hydr*ide*, magnesium brom*ide*. (*Note*: the metal cations are named first followed by the non-metal anions.)

When a chemist is trying to tell the difference between two ions of the same element, there are two ways to do it: the Stock system or the common naming system.

STOCK SYSTEM

The first, the *Stock system*, invented by German Chemist Alfred Stock, at Cornell University, uses Roman numerals to show which ion is which. When referring to lead (Pb^{2+}) you would say "lead two" and write it Pb(II).

In the Stock system, $FeCl_2$ would be called iron(II) chloride, $FeCl_3$ would be iron(III) chloride, and $PbBr_4$ would be called lead(IV) bromide. The IUPAC naming system uses the Stock method.

COMMON NAMING SYSTEM

The second method or *common naming system* uses the endings (suffixes) to distinguish between chemical forms. If an experiment calls for cuprous sulfate, it is talking about a copper Cu^+ ion. The ending "*ous*" is used to name the lesser charged ion of the different copper forms. The copper ion Cu^{2+} is known as a cupr*ic* ion, while Cu^+ is called the cupr*ous* ion, since it is the lesser charged ion of the two forms. The iron ion Fe^{3+} is called a ferr*ic* ion, while Fe^{2+} is known as a ferr*ous* ion.

Naming ionic compounds in chemistry is fairly simple when you remember to name the cation (+) first, followed by the anion (–). Writing formulas for ionic compounds is easy if you remember to keep the charges balanced. Then they almost name themselves.

In the compound $MgCl_2$ (magnesium chloride), one magnesium ion (Mg^{2+}) combines with two chloride ions (Cl^-). This combination makes the final molecule electrically neutral. It is written $MgCl_2$. The subscript 2 attached to the chloride shows that 2 ions of negatively charged chlorine ions are needed to balance the formula.

Simple math is involved in writing formulas of polyatomic ions, but it is not rocket science. Just take one step at a time.

> To find the **lowest common multiple (LCM)**, multiply each charge by whatever number works to give the least common multiple.

(*Tip: lowest common multiple is easiest to use when the charge from one ion is used as the multiplier for the other ion.*)

When writing the formula of gallium oxide, first write the element symbols, Ga^{3+} and O^{2-}. The charges are dropped when you write this. Look at the *LCM* of 3 and 2 or 6. To get 6 Ga and 6 O, you multiply

$$2\ (Ga^{3+}) + 3\ (O^{2-}) = 2\ (^{+}3) + 3(^{-}2) = 0$$

So the formula is written Ga_2O_3.

To write the formula for ammonium carbonate, $(NH_4)_2CO_3$, parentheses are used to show that two ions of ammonium are needed to balance one carbonate ion. (*Note*: two NH_4 ions balance one CO_3^{2-} ion.) Parentheses are used when writing the chemical formula of compounds of polyatomic ions. Parentheses are not needed unless a subscript is used. For the compound lead sulfate, $PbSO_4$, only one ion of lead is used, so no parentheses are needed.

Remember, to name a compound from its formula, write the name of the positive ion (cation) first, followed by the name of the anion (negative ion). So Li_2S is lithium sulfide.

In the case of some polyatomic ions, the naming of the different combining types can be remembered by checking the beginnings and endings of the element name. When there are only two types, the "ate" ending is used for the ion with more atoms. A sulfur atom may have four oxygen atoms around it (sulf*ate* ion, SO_4^{2-}) or three oxygen atoms (sulf*ite* ion, SO_3^{2-}). When four different types exist then the following rule applies:

- an ion with 4 other atoms is named: "per___ate"
- an ion with 3 other atoms is named: "___ate"
- an ion with 2 other atoms is named: "___ite"
- an ion with 1 other atom is named: "hypo___ite"

For polyatomic bromine ions, the naming would be *per*brom*ate* (BrO_4^{1-}), brom*ate* (BrO_3^{1-}), brom*ite* (BrO_2^{1-}), and *hypo*brom*ite*, (BrO^{1-}). **Figure 9.2** shows some clues to help with chemical naming.

The characteristics of different compounds are affected by their ionic character. It is easier to understand element bonding if you memorize or make flash cards to study these ionic forms and their types.

Covalent Compounds

Not all compounds are ionic. When non-metals react with other non-metals, covalently bonded compounds form. *Covalent compounds* are generally soft

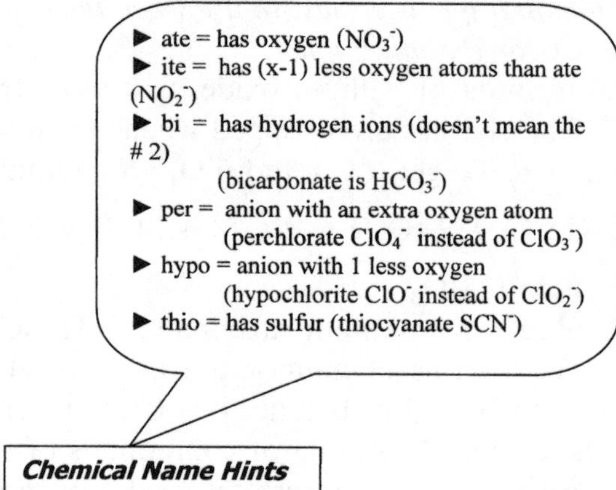

Fig. 9.2. A few clues to chemical naming can make bonding a lot easier to understand.

solids with low melting points. Many are liquids or gases at room temperature. When only two elements are bonded, the compound is called a *binary covalent compound.*

To name a binary covalent compound, the IUPAC method orders the non-metals in a certain way. This order lists which non-metals to name first:

$$B > Si > C > P > N > H > S > I > Br > Cl > O > \text{functional group}$$

In a compound containing phosphorus and oxygen, phosphorus is named before oxygen.

For binary compounds, chemical naming includes writing the number of atoms of each type of element. These are given by a Greek prefix. **Table 9.2** gives some Greek prefix tips to chemical naming.

To name a binary covalent compound, name the first non-metal, including its prefix, then name the second non-metal, using the prefix and changing the ending to "ide." (Note: the prefix *mono* is generally not used for the first non-metal in the formula.)

EXAMPLE 9.2

Name the following compounds: CO, CO_2, NO_2, N_2O_3, CCl_4, SF_6. Did you get carbon monoxide, carbon dioxide, nitrogen dioxide, dinitrogen trioxide, carbon tetrachloride, and sulfur hexafluoride?

With a good idea of how ions work, the characteristics and reactions of metals and non-metals in Chapter 12 will be a snap.

Table 9.2 Greek prefixes can make the counting of atoms easier when naming compounds.

Number	Greek prefix
1	Mono
2	Di
3	Tri
4	Tetra
5	Penta
6	Hexa
7	Hepta
8	Octa
9	Nona
10	Deca
12	Dodeca

Quiz 9

1. Monatomic ions
   (a) are pretty boring
   (b) have less than one oxygen
   (c) have only one atom
   (d) contain actinium

2. The chemical prefix bi means
   (a) two atoms
   (b) double the number of atoms

   (c) containing hydrogen
   (d) containing oxygen

3. Polyatomic ions
   (a) have more than one atom
   (b) less than one ion of oxygen
   (c) have only one type of atom
   (d) contain polonium

4. The chemical prefix hypo means
   (a) less than one nitrogen atom
   (b) triple the number of hydrogen atoms
   (c) contains sulfur
   (d) contains one less oxygen atom

5. Gaps were included in the first Periodic Table
   (a) because they didn't understand the octet rule
   (b) to allow for undiscovered elements
   (c) to make it easier to write on one page
   (d) to eliminate arguments about placement

6. Polyatomic ions
   (a) have no charge
   (b) have an overall charge
   (c) have ionic bonds
   (d) contain only oxygen

7. Iron can form how many ions?
   (a) 4
   (b) 3
   (c) 2
   (d) 1

8. Germanium was
   (a) discovered by a Swiss chemist
   (b) named from the Greek word for geranium
   (c) discovered in 1862
   (d) slid into an open gap in the first Periodic Table

9. Mendelevium with an atomic number of 101
   (a) has 101 protons
   (b) was named after the scientist Gregor Mendel
   (c) was discovered in 1855
   (d) has no isotopes

10. In chemistry, the lowest common multiple is
    (a) a negative factor used in general mathematics
    (b) easiest to use when the charge from one ion is used as the multiplier for the other ion
    (c) was first used by Albert Einstein
    (d) is only used in group IIA of the Periodic Table

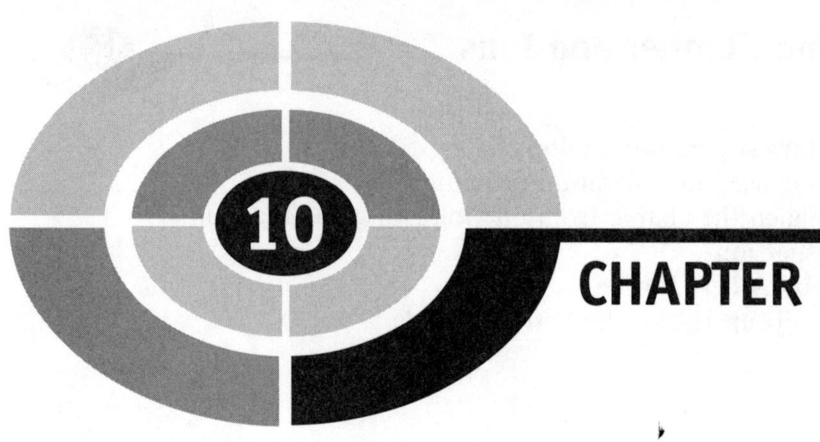

Organic Chemistry and Functional Groups

Organic chemistry, based originally on the study of living things like molds, plants, algae, red blood cells, gnats, and elephants, to name a few, is focused on compounds that include carbon. It is estimated that greater than 95% of all known chemicals contain carbon. At last count, there were over two million known organic compounds, nearly twenty times more than all the other known chemicals combined.

At one time, it was thought that organic compounds contained some type of "vital force" since they were once living organisms. However, when the organic compound urea (CH_4ON_2) was made in the laboratory in 1828 by Friedrich Wohler out of ammonium carbonate it was the first time an organic compound had been made from inorganic materials off the shelf. Chemists, then, started thinking about the possibilities of organic molecules and their reactions in a whole different way.

This largest group of covalently bonded compounds makes up the central study of petroleum-based chemicals, plastics, synthetic fibers, and biological chemistry. Petroleum, also known as crude oil, is made up of organic compounds from the decomposed remains of plants and animals that died millions of years ago.

> **Organic chemistry** is the chemistry of carbon.

The six electrons of carbon fill up the 1s, 2s, 2p orbitals. The four valence electrons tend to stay unpaired, allowing carbon to form four bonds. Carbon can form open chains, closed chains (rings), and a combination of open and closed chains.

The wide variety of carbon-containing compounds found in nature can be related to carbon's talent. Carbon is able to form long-chain molecules like decane ($C_{10}H_{22}$), branching macromolecules like natural rubber, and ring structures like menthol (from peppermint).

In the same group (IVA) as carbon, silicon is very much like carbon in atomic structure. It forms silicon–silicon covalent bonds, but since silicon is over double the size of carbon, the silicon bond lengths are longer and weaker. It is like a bridge between two river banks. The bridge across a 4 meter (12 foot) wide stream will be much stronger and more stable than one across an 8 meter (28 foot) stream, when the middle is not supported.

Figure 10.1 shows some common organic molecules.

Hydrocarbons

The broad group of organic compounds called *hydrocarbons* are made up of, you guessed it, molecules containing only carbon and hydrogen. Hydrocarbons are perhaps the easiest molecules in all of chemistry to learn. Once you get the basics, the rest is a matter of plugging in additional element groups.

Hydrocarbons are divided into subgroups depending on how carbon and hydrogen have bonded. Those hydrocarbons made up of only single bonds between carbon and hydrogen are known as *alkanes*. When carbon forms a double bond with hydrogen in a molecule, the subgroup is known as *alkenes*. Similarly, molecules with triple bonds between carbon and hydrogen are known as *alkynes*. The two simplest members of the alkane group are methane and ethane.

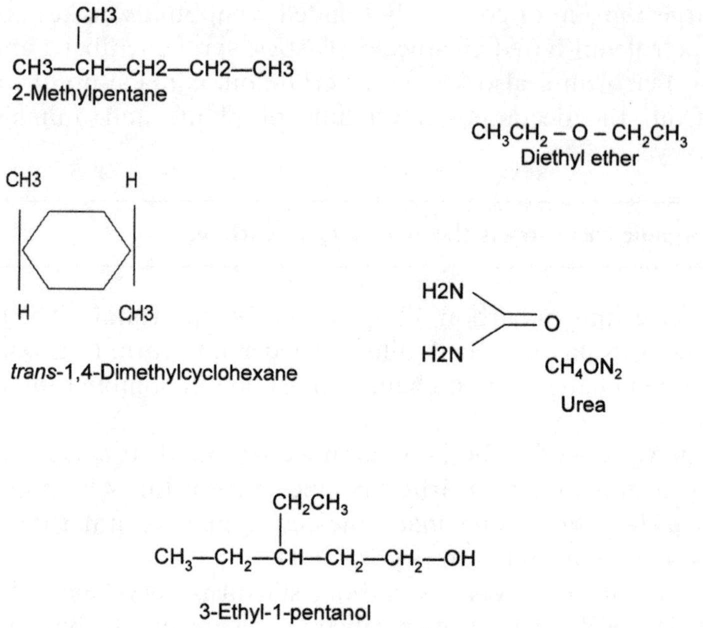

CH3—CH(CH3)—CH2—CH2—CH3
2-Methylpentane

$CH_3CH_2 - O - CH_2CH_3$
Diethyl ether

trans-1,4-Dimethylcyclohexane

H2N(H2N)C=O
CH_4ON_2
Urea

CH_3—CH_2—CH(CH_2CH_3)—CH_2—CH_2—OH
3-Ethyl-1-pentanol

Fig. 10.1. These organic molecules give you an idea of the diversity possible in organic chemistry.

Complete carbon bonding is seen in methane (CH_4). Carbon forms a tetrahedral (four-sided) compound with hydrogen, sharing electrons at bond angles of about 109 degrees. **Figure 10.2** shows methane with bond angles shown.

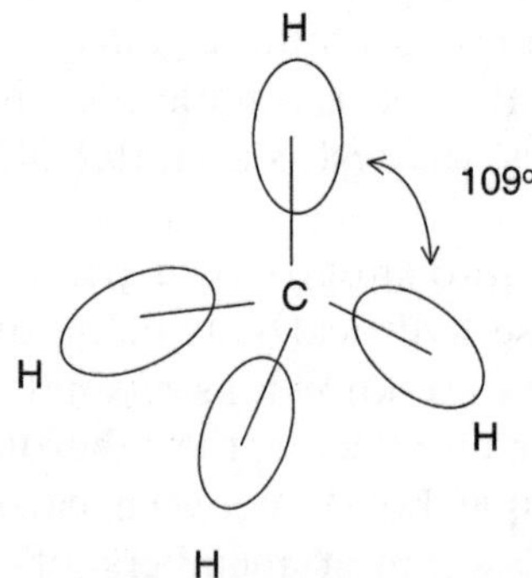

Fig. 10.2. The bond angles of 109° each are shown in the methane molecule.

Substitution Reactions

Hydrocarbons react with other elements in specific ways. The bond to the hydrogen is broken and the reacting element is slipped into its place. For example, methane's reaction with members of the halogen group (fluorine, chlorine, or bromine) produce halomethanes and a hydrogen halide.

EXAMPLE 10.1
The substitution reactions that occur between a hydrocarbon (methane, CH_4) and a halogen (chlorine, Cl) are shown.

$$CH_4 + Cl_2 \Rightarrow CH_3Cl + HCl$$

The same reaction takes place with additions of the halogen to the singly substituted CH_3Cl.

$$CH_3Cl + Cl_2 \Rightarrow CH_2Cl_2 + HCl$$

$$CH_2Cl_2 + Cl_2 \Rightarrow CHCl_3 + HCl$$

$$CHCl_3 + Cl_2 \Rightarrow CCl_4 + HCl$$

Substitution reactions are simple and very common among organic compounds, but they can occur between other compounds too.

EXAMPLE 10.2
A substitution reaction between fluoromethane (CH_3F) and potassium bromide (KBr) trades elements in the following reaction:

$$CH_3F + KBr \Rightarrow CH_3Br + KF$$

Naming of Alkanes

The International Union of Pure and Applied Chemistry (IUPAC) method of naming alkanes is fairly simple. The names of all alkanes end in *ane*. The prefixes, based on the number of carbons in the molecule, make naming a lot like counting. Meth- (1 carbon), eth- (2), prop- (3), but- (4), pent- (5), hex- (6), hep- (7), oct- (8), non- (9), and dec- (10) are the basic prefixes. If you know these, you will be able to breeze through naming.

EXAMPLE 10.3
What are the names of the following alkanes?

(1) CH_4, (2) C_3H_8, (3) C_6H_{14}, and (4) C_9H_{20}.

Did you get (1) methane, (2) propane, (3) hexane, and (4) nonane?

The names of functional groups resulting from alkanes after the loss of a hydrogen are called *alkyl* groups. Methane, then, becomes *methyl*, ethane becomes *ethyl*, and so on.

Double Bonds

The naming of alkenes or double-bonded molecules is very much like that of the alkanes.

Alkenes are numbered using the following IUPAC rules:

(a) The base molecule name comes from the longest chain that contains the double bond. (So if the longest chain has 4 carbons, the base name would be butene.)
(b) The chain is numbered to include both carbons of the double bond. (Then $CH_2{=}CHCH_2CH_3$ is 1-butene.)
(c) The locations of the attached groups are numbered as to the carbon to which they are attached. (So 2-methyl-2-butene has a methyl group ($-CH_3$) on the second carbon of the double bonded-butene.)
(d) When naming 6-carbon ring molecules, the carbons are numbered clockwise around the ring. (So 2-methylbenzene has a methyl (CH_3) attached at the second carbon in the ring.)

Ethene (C_2H_4) is an example of a simple double-bonded carbon molecule, with two hydrogen atoms bonding to each carbon and the two carbons connected by a double bond. This is really only part of the answer. In a double covalent bond, two pairs of electrons are shared between two atoms instead of one pair. **Figure 10.3** shows how the electrons are shared in ethene.

Double bonds also hold molecules into rigid shapes, since there is no rotation or twisting possible at the double bond. Therefore, a general rule-of-thumb is that a molecule that does not possess a double or triple bond is more likely to be able to roll up or twist into different structures more easily. **Figure 10.4** shows some common examples of double bonds.

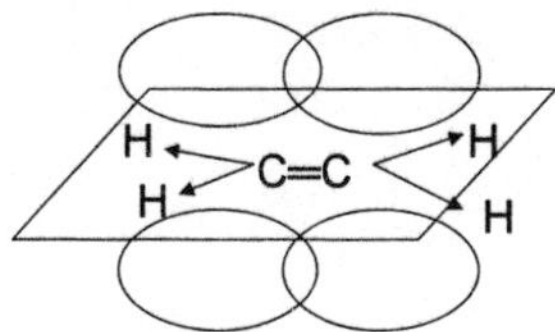

Fig. 10.3. The electrons are shared in the ethene molecule in a flat plane configuration.

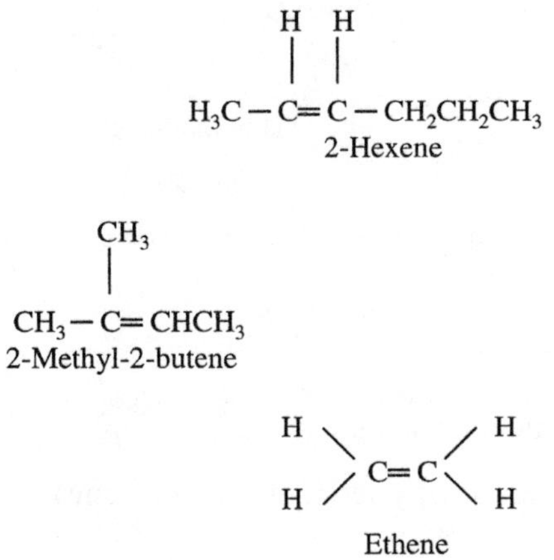

$$H_3C-\underset{\text{}}{\overset{H}{\overset{|}{C}}}=\overset{H}{\overset{|}{C}}-CH_2CH_2CH_3$$

2-Hexene

$$CH_3-\overset{CH_3}{\overset{|}{C}}=CHCH_3$$

2-Methyl-2-butene

$$\begin{matrix} H & & & & H \\ & \diagdown & & \diagup & \\ & & C=C & & \\ & \diagup & & \diagdown & \\ H & & & & H \end{matrix}$$

Ethene

Fig. 10.4. The double bond can have different elements and groups attached to the carbon.

Six-sided ring structures, like benzene, with alternating single and double bonds are called *aromatic* hydrocarbons. Many of these compounds have smells that most of us know. Oil of vanilla or *vanillin* ($C_8H_8O_3$), extracted from the fermented seed pods of the vanilla orchid, is shown in **Figure 10.5**. Another common aromatic compound, *cinnemaldehyde* or oil of cinnamon (C_9H_8O), obtained from the steam distillation of cinnamon tree bark, is also shown.

Triple Bonds

When carbon forms a triple bond with another carbon, it is called an *alkyne*. Originally, triple-bonded carbon compounds were named after the simplest molecule, acetylene (C_2H_2), which is $HC{\equiv}CH$. Dimethylacetylene is $CH_3C{\equiv}CCH_3$.

$C_8H_8O_3$
Vanilla

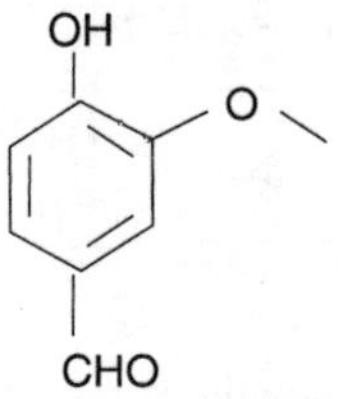

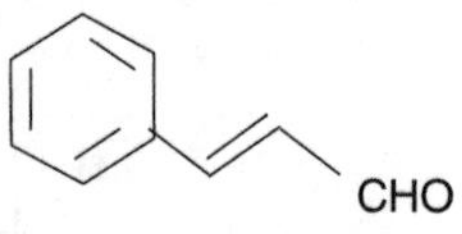

Oil of Cinnamon
C_9H_8O

Fig. 10.5. Oil of cinnamon and oil of vanilla are well-known organic compounds.

EXAMPLE 10.4

Alkynes are also known as acetylenes such as:

$$CH_3C{\equiv}CH, \text{ propyne (methylacetylene)}$$

or

$$CH_3CH_2C{\equiv}CH, \text{ 1-butyne (ethylacetylene)}$$

Triple bonds involve the sharing of three pairs of electrons or six electrons. **Figure 10.6** shows some alkyne molecules.

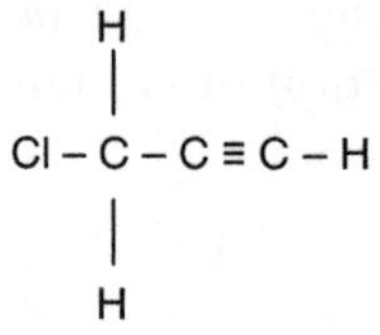

1-Chloro-2-butyne

H-C ≡ C-H
Ethyne
(acetylene)

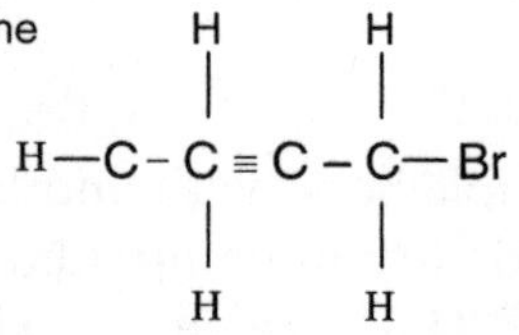

1-Bromo-2-butyne

Fig. 10.6. Alkynes are triple-bonded carbon compounds.

Ring Structures

When carbons are arranged at the corners of a hexagon with a hydrogen bonded to each carbon and alternating double bonds between carbons, it is known as a ring structure or aromatic ring. The most basic ring structure is that of benzene (C_6H_6).

Functional groups that are substituted for hydrogens around a ring bond to carbons and produce a variety of different molecules.

Bond Polarity

In covalent bonds, when electrons are shared by the same elements, the electron distribution is symmetrical. If the atoms of two or more different elements share a pair of electrons, the electron density at one or another of the atoms will be different. Some atoms have a stronger attraction for a shared electron pair than another. The bonding is not equal in the general sharing (one atom gets more time with the electron pair than the other). This unequal sharing causes a slight charge shift to one atom more than the other.

> **Bond polarity** occurs when electron pairs are unequally shared between atoms of different elements.

It is important to keep in mind that some molecules may have polar bonds, without molecular polarity, due to the canceling effects of symmetry. In carbon tetrachloride (CCl_4) there are four polar bonds, but the bond angles and symmetry causes the bond dipoles to cancel out. Carbon tetrachloride, then, is a non-polar molecule.

This unequal sharing or stronger/weaker attraction of shared electrons explains why some molecules have certain reactivity with some atoms and very different reactivity with others. Melting and boiling points as well as the ability to form higher energy compounds depends on these polarization effects.

Addition Reactions

When compounds contain double and triple bonds, the most common element interactions take place as *addition reactions*.

When a double-bonded hydrocarbon like ethene ($CH_2{=}CH_2$) is added to another compound with single bonds, the elements of the new compound position themselves to either side of the double bond. The double bond is broken and single bonds are formed.

EXAMPLE 10.5

The addition of hydrogen chloride to ethene yields chloroethane:

$$CH_2{=}CH_2 + HCl \Rightarrow H_3\text{–}C\text{–}C\text{–}H_2Cl \Rightarrow CH_3CH_2Cl$$

When diatomic chloride is added to ethene, then the result is 1,2-dichloroethane:

$$CH_2{=}CH_2 + Cl_2 \Rightarrow ClH_2\text{–}C\text{–}C\text{–}H_2Cl \Rightarrow CH_2CH_2Cl_2$$

For triple-bonded hydrocarbons, the addition of bromine to ethyne would yield:

$$H\text{–}C{\equiv}C\text{–}H + 2\ Br_2 \Rightarrow Br_2H\text{–}C\text{–}C\text{–}HBr_2 \Rightarrow CH_2Br_4$$

Functional Groups

In organic chemistry, carbon combines with oxygen and hydrogen in specific ways to produce standard functional groups such as alcohols, phenols, and carboxylic acids (–OH); aldehydes, ketones, esters, and carboxylic acids (C=O); amines (N–H); and nitriles (C≡N), which then react in known ways.

Some of these organic molecules are long and complex. To save time and space, certain *functional groups* or specific groupings of elements that do a certain job are written in chemical shorthand.

To highlight added functional groups, R is commonly used to stand for the carbon group CH_3–, CH_3CH_2–, $CH_3CH_2CH_2$–, or CH_3CHCH_3–.

EXAMPLE 10.6

The following shorthand formulas show how a few functional groups are added to the base carbon group (R):

- Ethanol is CH_3CH_2OH or R–OH
- Acetic acid (carboxylic acid, commonly known as vinegar) is CH_3CO_2H or R–COOH
- Formaldehyde (methanal) is R–HC=O
- Methylamine is CH_3NH_2 or R–NH_2
- Benzylchloride is a benzene ring–CH_2Cl.

Figure 10.7 shows some functional groups that combine with carbon to make organic molecules.

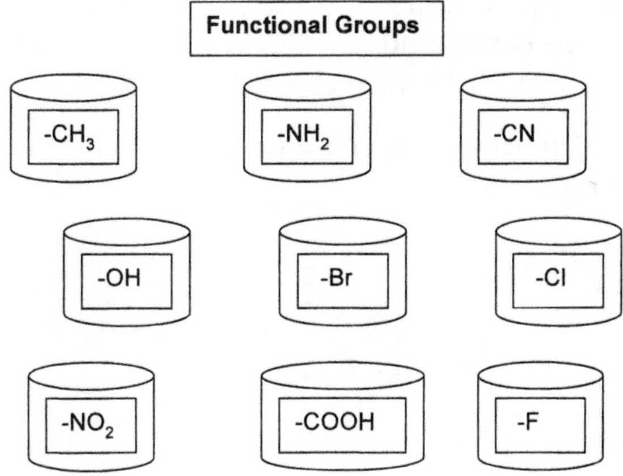

Fig. 10.7. A wide variety of functional groups combine with carbon to produce organic molecules.

These functional groups make deciphering organic reactions much simpler. Finding functional groups on a long carbon chain (R) makes identifying the core molecule easier. Once the central compound is known, the other reactants can be recognized.

Homologous Series

A *homologous series* of organic compounds are those that are very much alike structurally or react pretty much the same. The formula is the same for the compounds in the group with a CH_2 group added.

The alkanes are a homologous series of compounds. Their general formula expands each time by a –CH_2 group and is C_nH_{2n+2}. Since they have only single covalent bonds, they are known as *saturated* compounds. Every carbon is bonded to four other atoms or groups.

The importance of organic molecules and their characteristics will become clearer when we study chemical bonding in Chapter 13.

Quiz 10

1. Organic chemistry is based on
   (a) nitrogen compounds
   (b) protein polymerization
   (c) carbon-based compounds
   (d) the theory of particle-wave chemistry

2. Which of the following does not contain carbon?
   (a) carbohydrates
   (b) jet fuel
   (c) synthetic fibers
   (d) table salt

3. The simplest hydrocarbon molecule is
   (a) methane
   (b) ethane
   (c) propane
   (d) butane

4. The bond between the carbons in ethene is a
   (a) single bond
   (b) triple bond
   (c) double bond
   (d) quadruple bond

5. Tetrahedral bonding angles of carbon are about
   (a) 20 degrees
   (b) 45 degrees
   (c) 90 degrees
   (d) 109 degrees

6. Alkanes are
   (a) formed from NH_2 groups
   (b) an example of a homologous series
   (c) composed of many different element groups
   (d) only found in inorganic compound reactions

7. Since ethyne is a linear molecule, the two carbons form a triple bond of
   (a) 180 degree angles
   (b) 45 degree angles
   (c) 120 degree angles
   (d) 109 degree angles

8. Bond polarity
   (a) occurs in the northern hemisphere of the globe
   (b) occurs when electron pairs are unequally shared between atoms
   (c) was discovered in 1862
   (d) is best seen in group VIII of the Periodic Table

9. Saturated hydrocarbon compounds
   (a) contain many double bonds in long chains
   (b) contain triple bonds to oxygen molecules
   (c) are open to hydrogen bonding
   (d) contain only single bonds

10. Carbon is
    (a) the element that bonds with sulfur to form cyanide
    (b) unable to form triple bonds with nitrogen
    (c) found in greater than 95% of all known chemicals
    (d) number 8 in the Periodic Table and has an atomic weight of 16

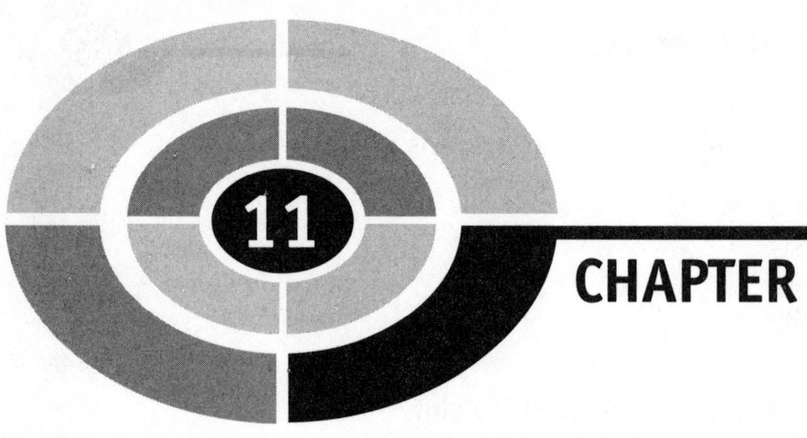

Radiochemistry

The nucleus of an element contains protons and neutrons. We have already learned that an element's identity and atomic number is found from the number of protons in a nucleus. However, around 1912, researchers found odd changes in samples of the same elements. Their atomic number was the same, but they had different atomic masses. Chemists couldn't decide if they had discovered new elements or different forms of old ones.

Isotopes

In 1913, Frederick Soddy named these chemically identical elements with different atomic weights *isotopes*, from the Greek word meaning "same place," since they were placed in the sample's same spot on the Periodic Table.

> **Isotopes** are chemically identical atoms of the same element but with different numbers of neutrons and different mass numbers.

The naturally occurring isotope of hydrogen, called *deuterium*, was discovered by Harold Urey in 1931. Urey found that certain samples of hydrogen were twice the weight of common hydrogen. His experiments on simulating early Earth's atoms earned him the Nobel Prize for Chemistry in 1934.

Although most hydrogen atoms have a nucleus of only one proton and no neutrons, 1 out of every 5,000 hydrogen atoms have a nucleus of 1 proton and 1 neutron (deuterium). When common hydrogen (also known as protium) is measured, then, it is found to have a mass of one, while deuterium atoms have a mass of two. The even rarer radioactive isotope of hydrogen called *tritium* has a mass of three, with 1 proton and 2 neutrons. **Figure 11.1** illustrates the three forms of hydrogen.

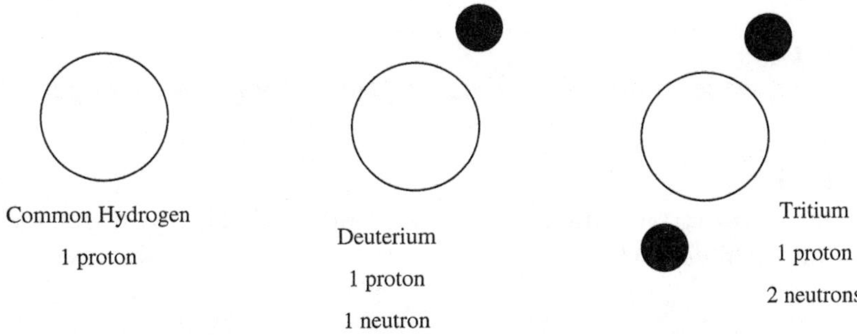

Fig. 11.1. Hydrogen has three different forms that are named separately.

It is interesting that deuterium and common hydrogen both react with oxygen to form water. Regular water has a mass of 18 grams, while "heavy water" has a mass of 20. Though less likely to form the same compounds as hydrogen and deuterium, tritium does enter into reactions.

Naming Isotopes

Hydrogen is the only element that has been given separate names for its different isotopes, perhaps because it is involved in so many different kinds of reactions.

In naming isotope forms of elements other than hydrogen, you basically have two methods to choose from. One way is to write the element name with a hyphen and then the mass number. The second isotope shorthand method uses the element symbol along with the atomic number (Z) as a subscript *and* the mass number as a superscript (A). Both are written to the left side of the element's symbol. (*Note*: atomic mass is not the same as atomic number.) **Figure 11.2** shows both ways of writing isotopes for radioactive radon, which has over 20 different isotopes. Radon-222 is the longest-lived of these isotopes with a half-life of nearly four days.

Rn-222

$^{222}{}_{86}Rn$

Fig. 11.2. There are two different ways to name radioactive isotopes. Radioactive radon is shown.

EXAMPLE 11.1
What are the atomic numbers of ^{90}Sr, ^{37}Cl, and ^{24}Mg? Did you get 38, 17, and 12?

EXAMPLE 11.2
Can you name these elements: $^{12}_{6}X$, $^{238}_{92}X$, and $^{99}_{43}X$? Did you get carbon, uranium, and technetium?

Radioactivity

Radioactivity of chemical elements, sometimes called *radiochemistry*, was discovered in 1896 by Antoine Becquerel, when he found that a photographic plate never exposed to sunlight in his lab had become exposed. The only possible culprit was a nearby uranium salt sitting on the bench top.

The term *radioactivity* was first used by French scientist Marie Curie in 1898. Marie Curie and her physicist husband, Pierre, found that radioactive particles were emitted as either electrically negative (–) which were called beta (β) particles or positive (+) called alpha (α) particles.

The first years following the discovery of the special properties of radioactive elements, with all of their wonders, led Pierre and Marie Curie to plan more studies. In 1903, Becquerel and the Curies shared the Nobel Prize for Physics for their work in radioactivity.

In 1911, after the discovery of polonium and radium the year before, the Curies received another Nobel Prize, this time for chemistry for their continued work.

Nuclear Reactions

Most chemical reactions are focused on the outer electrons of an element, sharing, swapping, and bumping electrons into and out of the combining partners of a reaction. Nuclear reactions are different. They take place within the nucleus.

There are two types of nuclear reactions. The first is the radioactive decay of bonds within the nucleus that emit radiation when broken. The second is the "billiard ball" type of reaction where the nucleus, or a nuclear particle (like a proton), is struck by another nucleus or nuclear particle. It is easy to remember these with the following: element → decay → radiation (see **Figure 11.3**).

Element→Decay →Radiation

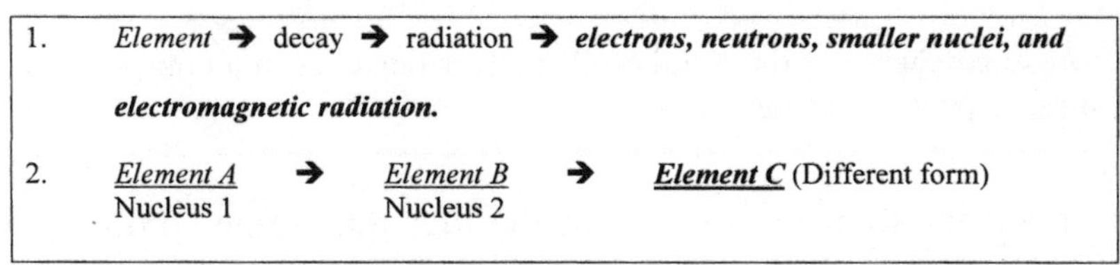

Fig. 11.3. Radioactive elements decay and lose energy in an ordered way.

Radioactive Decay

A radioactive element, like everything else in life, decays (ages). When uranium or plutonium decays over billions of years, they go through a transformation process of degrading into lower energy element forms until they settle into one that is stable.

When a radioactive element decays, different nuclear particles are given off. These radiation particles can be separated by an electric (magnetic field) and detected in the laboratory:

$$\text{beta } (\beta) \text{ particles} = \text{negatively } (-) \text{ charged particles}$$

$$\text{alpha } (\alpha) \text{ particles} = \text{positively } (+) \text{ charged particles}$$

Gamma (γ) particles are electromagnetic radiation with no overall charge, similar to X-rays but with a shorter wavelength.

Decay of radioactive isotopes is affected by the stability of an element at a certain energy level. Bismuth (Bi), at atomic number 83, is the heaviest element in the Periodic Table with a minimum of one stable isotope. All other heavier elements are radioactive.

Magic Numbers

Radioactive decay seems to be a mathematical process. Experimenters have found that protons have the magic numbers of 2, 8, 20, 28, 50, 82, and 114. Neutrons have the same magic numbers as protons, plus the number 126. Radioactive uranium (^{238}U) decays eventually to lead (^{82}Pb).

> **Magic numbers** are the number of nuclear particles in a completed shell of protons or neutrons.

Magic numbers seem to come from the fact that some combinations of protons and neutrons are stable (non-radioactive), while some are unstable (radioactive). The radioactive elements decay (lose energy) over time. During decay, the unstable nucleus tries to become more stable by emitting particles. This goes on in a stepwise manner until a final, stable configuration of protons and neutrons is achieved.

Half-life

All radioactive isotopes have a specific, set *half-life*. These time periods are not dependent on pressure, temperature, or bonding properties.

> The **half-life** of a radioactive isotope is the time needed for half of an elemental sample to decay.

For example, the half-life of $^{238}_{92}U$ is 4.5×10^9 years. This is about the same as the age of the Earth. It is amazing to think that the uranium found today will be around for another four billion years.

Nuclear Bombardment (Billiard Ball Reaction)

In 1919, Ernest Rutherford discovered he could turn one element into another element (change from its original energy level to the energy level of a different element) by colliding the nucleus of one element with the nucleus of another.

> **Transmutation** is the process of one element changing into another element.

Transmutation occurs when one element changes into a different element by hitting the nucleus of the first element with nuclear particles (protons, neutrons) of another element and changing its nuclear content.

Rutherford used alpha particles to collide with nitrogen nuclei in the laboratory. Protons were thrown off in the reaction. The equation $^{14}N + {}^{4}He \rightarrow {}^{17}O + {}^{1}H$ shows this process. Currently, *particle accelerators* are used to speed up electrons, protons, and alpha particles to super speeds. These super speeds are needed to penetrate the nucleus of elements with large atomic numbers.

Think of it as tossing a tranquilizer dart at a rhinoceros. The rhino's tough hide would cause the dart to bounce off. Very high speeds are needed to penetrate. Slower speeds would have little effect. In the same way, nuclear particles must be greatly accelerated to penetrate the tight core of another element's nucleus.

Transuranium Elements

Elements with atomic numbers greater than uranium (^{92}U), the element with the highest atomic number found in nature, are called transuranium elements. All the transuranium elements of the actinide series were discovered as synthetic radioactive isotopes at the University of California at Berkeley or at Argonne National Laboratory. By colliding uranium (^{238}U) with neutrons, they produced uranium (^{239}U), which days later decayed to neptunium (^{237}Np).

Americium and curium were placed after actinium in the actinide series in the Periodic Table because their chemical properties were similar. Since that time, elements with atomic numbers to 118 have been reported by scientists around the world.

Radioactivity is often thought of as a terrible side effect of nuclear weapons and X-rays, but when properly shielded, radioactive elements have a variety of uses. The actinide series of elements are used as power supplies for pacemakers, nuclear satellites, and submarines. Americium–241 is used in home smoke detectors to control the conductivity of air that changes when smoke is present.

Radiation Detectors

There are two types of equipment used to detect radiation, *ionization counters* and *scintillation counters*.

> **Ionization counters** detect the production of ions in matter.

The Geiger counter, invented in 1928 and named after one of its two inventors, H. Geiger and E.W. Muller, counts particles emitted by radioactive nuclei in a non-reactive noble gas, like argon. Alpha and beta particles are detected this way.

> **Scintillation counters** detect nuclear radiation from flashes of light made when radiation affects a sample.

A sample that emits flashes of light when struck by radiation is called a *phosphor*. Rutherford used zinc sulfide as a phosphor to detect alpha particles. Sodium iodide and thallium combined together are used to detect gamma radiation in the same way.

Carbon Dating

Isotopes are also used in the dating of ancient soils, plants, animals, and the tools of early peoples. The isotope of carbon ^{14}C, which has a half-life of 5,730 years, can be used to calculate the age of old things. Since the rate of radioactive decay is constant, observing the decay rate allows the calculation of years that have past in relation to the half-life.

Carbon dating of organic materials depends a lot on the preservation of the original sample. Most people agree that carbon dating is only accurate to between 30,000 and 50,000 years. Other radiometric methods that make use of uranium, lead, potassium, and argon can measure much longer time periods since they are not restricted to organic samples that were once alive. The oldest rocks tested on earth have been dated around four billion years old. This radioactive dating allowed scientists to measure the earth's crust at near this age and meteorites at around 4.4×10^9 years.

Nuclear Medicine

Nuclear medicine is the specific focus of radiology that uses minute amounts of radioactive materials, or *radiopharmaceuticals*, to study organ function and structure. Nuclear medicine imaging is a combination of many different sciences, including chemistry, physics, mathematics, computer technology, and medicine.

Because X-rays travel through soft tissue, such as blood vessels, heart muscle, and intestines, contrast tracers are used in nuclear imaging. Nuclear imaging examines organ function and structure, whereas diagnostic radiology is based on the anatomy of bones and hard tissues.

A relatively new use for radioactive isotopes/radiopharmaceuticals is in the detection and treatment of cancer. This branch of radiology is often used to help diagnose and treat abnormalities very early in the stages of a disease, such as thyroid cancer. Since rapidly dividing cells like those present in cancer are more vulnerable to radiation than slow-growing normal cells, treatment using medical isotopes works well.

Radon-222, used initially, and now cobalt-60, are used as implants near a cancer or shot as a narrow beam to an inoperable brain cancer or used before surgery to shrink a tumor in lung or breast cancers.

A *radioactive tracer* is a very small amount of radioactive isotope added to a chemical, biological, or physical system in order to study the system. For example, radioactive barium (^{37}Ba) is used to diagnose unusual abdominal pain, gastroesophageal reflux, and gastric or duodenal ulcers or cancer. Thallium (^{201}Tl) is a radioactive tracer used to detect heart disease. This isotope binds with great enthusiasm to heart muscle that is well oxygenated. When a patient with heart trouble is tested, a scintillation counter detects the levels of radioactive thallium that have bonded to oxygen. When areas of the heart are not receiving oxygen, there is very little thallium binding and it is seen as a dark area.

Radioactively tagged enzymes, or proteins containing radioactive components, are used in the medical area of study called *radioimmunology*. Radioimmunology measures the levels of biological factors (proteins, enzymes) known to be changed by different diseases. Some of these medical isotopes used in medicine include phosphorus (^{32}P), iron (^{59}Fe), and iodine (^{131}I).

Radioactive elements are an important area of chemical study. New uses and methods of detection will increase this area of research as more and more nuclear sub-particles are discovered.

In Chapter 18 we will look at cutting-edge research in this area.

Quiz 11

1. Radioactivity is best described as
   (a) a radio signal accompanied by energy flow
   (b) violent reaction with water
   (c) extremely low levels of visible energy release
   (d) spontaneous disintegration of isotopes and radiation emission

2. Frederick Soddy named isotopes from the Greek word *iso* meaning
   (a) glowing
   (b) same place
   (c) easily seen
   (d) brittle

3. Elements with the same number of protons, but a different number of neutrons are called
   (a) alkali metals
   (b) electron receptors
   (c) isotopes
   (d) covalent bonds

4. Beta (β) particles are
   (a) ions of neutral charge
   (b) negatively (–) charged particles
   (c) positively (+) charged particles
   (d) non-existent in the universe

5. Tritium has
   (a) 1 electron and 1 proton
   (b) 1 electron and 1 neutron
   (c) 2 protons and 2 electrons
   (d) 1 proton and 2 neutrons

6. What is "heavy" water?
   (a) water with mercury added
   (b) water made mostly from deuterium and oxygen
   (c) water made from 2 molecules of hydrogen
   (d) an unreactive, neutral form of distilled water

7. What is the atomic number of ^{24}Mg?
   (a) 8
   (b) 12
   (c) 22
   (d) 37

8. Radioactive decay generally
   (a) occurs within seconds
   (b) gives off a great amount of heat
   (c) takes place as a transforming process over years
   (d) all of the above

9. Which of the following are all magic numbers?
   (a) 2, 8, 20, 28, 50, 82, 114
   (b) 2, 8, 20, 26, 82
   (c) 2, 8, 12, 22, 50, 82, 114
   (d) 2, 8, 18, 24, 48, 60, 82

10. Transmutation occurs when the
   (a) electrons are thrown off in a reaction
   (b) low-speed interaction occurs between metals
   (c) neutrons have the same magic number as electrons
   (d) nucleus of an element is hit by particles of another element

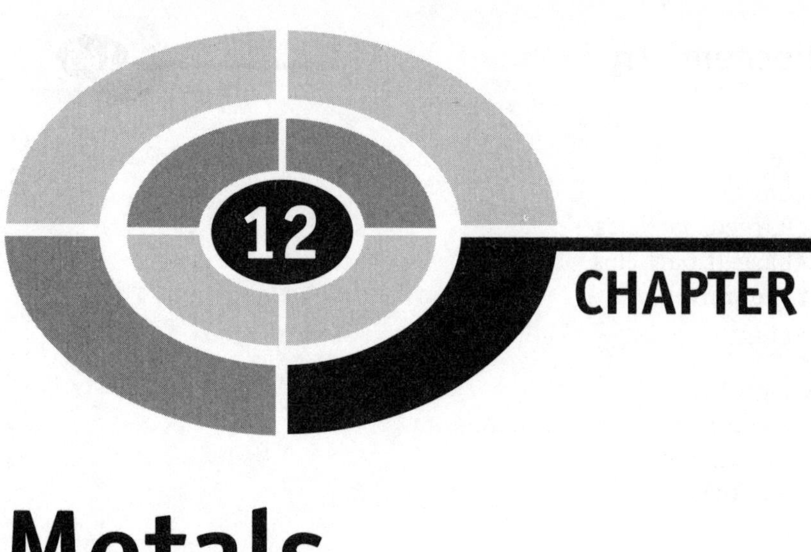

Metals

For thousands of years, humans have used metals as tools and weapons. They experimented with many native ores to make the best use of what they had. Of all the elements in the Periodic Table, the best known are the metals.

Ancient peoples found that when a very soft metal, like copper, was mixed with zinc and nickel, a harder and stronger metal, called bronze, was formed. This metal was used so much for cutting tools and other utensils that the years between 4,000 and 1,100 BC were called the Bronze Age.

Metals

We have all seen pictures of gold miners who have worked for years in dimly lit mines searching for deposits of gold that would "make their fortunes." Others spent their days from morning until night hunched over icy mountain streams panning handfuls of dirt and gravel and watching for the bright glint of a single metal nugget. Even today, people get excited over flecks of gold (iron sulfide, FeS_2) found in rock. Elements are commonly divided between either metals or non-metals.

Ores

Unlike gold and silver, which are pure elements, many metals are not found in nature as a single element. Most metals are combined with other elements within ores that must be processed to extract their different parts. **Table 12.1** gives some examples of different ores and the metals they contain.

Table 12.1 Mineral-rich ores contain a combination of two or more elements.

Element	Ore	Found in
Aluminum	Bauxite	France, Jamaica
Bismuth	Bismite	USA
Chromium	Chromite	South Africa, Russia
Cobalt	Cobaltite	Germany, Egypt
Copper	Chalcopyrite	Cyprus, USA, Canada
Iron	Hematite	USA, Australia
Lead	Galena	USA, Brazil, Canada
Mercury	Cinnabar	Algeria, Spain
Nickel	Pentlandite	Canada
Tin	Cassiterite	Bolivia
Titanium	Ilmenite	Peru, Kenya
Tungsten	Wolframite, scheelite	Spain, China
Zinc	Sphalerite	Australia, Canada, USA

Pure metals are separated from ores primarily with heat. This is done in a high-temperature blast furnace. By adding reactants like limestone and coke (a carbon residue) to break hydrogen bonding and release the bonded metals, individual metals can be collected. **Figure 12.1** shows a simple blast furnace.

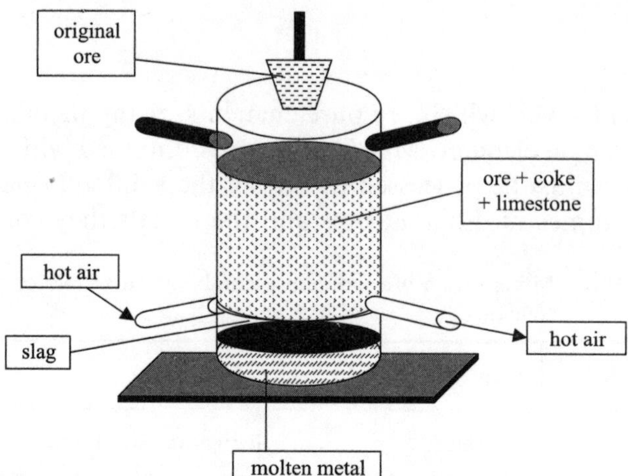

Fig. 12.1. A simple blast furnace is used to extract metals from the ores that contain them.

Lead, though sometimes found as a pure metal in nature, is usually found as the ore galena or lead sulfide. Lead ore is crushed, heated in a blast furnace, and then extracted. Most lead produced in the United States is used for battery and battery electrodes as well as lead solder used in making connections on computer circuit boards.

Mercury is most often found in nature as the ore, cinnabar. Cinnabar (HgS), also called vermillion, is a bright red mineral crushed and used to make the red paint pigment used by Renaissance painters.

The metal bismuth is most often found as the ores bismite (bismuth oxide, a yellow pigment in cosmetics) and bismuth glance (bismuth sulfide). It is commonly combined with lead, tin, and copper and so is extracted along with these metals. Bismuth is used in the treatment of ulcers by acting as an antacid. Like water, bismuth expands when it changes from a liquid to a solid.

Noble Metals: Gold, Silver, and Copper

Since the first shiny speck caught the eye of early humans, gold, silver, and copper have been used for coins, jewelry, and household serving ware. Resistant to rust and corrosion, they were an excellent choice for coins, while being doubly useful by showing the local king's face to strangers passing through the country.

Gold, a shiny yellow metal, is a good conductor of heat and electricity. It is the most malleable and ductile metal. The early alchemists based their reputations and lives on providing more of this metal to their patrons. In the western United States, gold fever affected thousands of people during the Gold Rush days of the 1800s seeking their fortunes and a better life.

Silver, a brilliant white, lustrous metal, is the best conductor of heat and electricity of all the metals. It was also prized by early peoples for its beauty and uses. Silver, less resistant to corrosion, will tarnish, turning black when it oxidizes in the air. It was thought that the state of Nevada was admitted to the Union in 1864, during the Civil War to provide funds to the Union and easier access to its resources of silver. This useful modern metal is used in coins, jewelry, electrical contacts, mirrors, circuitry, photography, and batteries.

Copper has an orange-brown color and is used in pipes, electrical wires, coins, paints, fungicides, and in alloys combined with other metals. In less developed countries, local people used copper widely for platters bowls, tools, and jewelry. Pennies, though once 100% copper, are now (since 1981) only treated on the outside with copper plating to give the United States' one cent coin its reddish brown (copper) color. Many years ago, the badges of police officers were made from copper and so the slang expression "copper" or "cop" was commonly used.

Alkali Metals

Within the alkali metals, lithium, sodium, and potassium are known as *active metals*. Their reactivity increases as their atomic number increases. These elements are extremely reactive in water and air. They are frequently stored in oil to prevent explosions when accidentally mixed with water.

Lithium is the third element in the Periodic Table following helium. It is usually found in the mineral spodumene. Lithium is the lightest metal and, when isolated, so soft it can be cut with a sharp knife. The density of lithium is so low that it will float on water when laid gently on the surface. As with the rest of the alkali metals, it is very reactive in water. Generally, it is stored in oil or kerosene to prevent it from reacting. When combined with aluminum, lithium combines to form a strong alloy metal used in aircraft and spaceships.

Although they are all highly reactive, the alkali metals all have certain differences. Sodium and potassium are soft metals that are found in silicate minerals and in seawater. Rubidium and cesium are the largest and heaviest elements of the group that react explosively with water.

Like the rest of the alkali metals, cesium is silvery white when in purified form. It is commonly found in the mineral pollucite, a compound containing silicon, oxygen, and aluminum as well. Cesium is the softest metal known and melts at 29°C. When held, it will melt at body temperature, 37°C (98.6°F), like a chocolate candy in your hand. Only mercury has a lower melting point.

Alkaline Earth Metals

The Periodic Table is great for placing elements according to their characteristics. An element's location provides information about its "personality" and uses.

Barium's location in group IIA shows that it is an alkaline earth metal. Like calcium and magnesium, barium has applications in medicine as barium sulfate (opaque to X-rays and used to check out the digestive tract) and photography (a whitener in photographic papers). Barium helps doctors to differentiate between physiological structures.

Metallic Crystals

Metals form large crystalline structures with high boiling and melting points. These structures are made up of metal ions. The extra electrons within the outer shells of metal atoms are still able to move around within the crystalline structures and in turn, cause the solids to be good conductors.

You can think of the structure as being like a vegetable soup, with the "broth" made up of the electrons, and the "vegetables" consisting of a lattice of positively charged metal ions. The broth does not have enough electrons to form individual bonds between atoms, so sharing electrons is a much more efficient form of bonding. The bonding is stronger and the metallic crystal is harder, as the ions are held more tightly between atoms that have more outer orbital electrons to share in the "broth." Transition metals with more electrons in their outer shell orbits are denser than alkali metals with fewer electrons in their outer shells.

Iron

Next in the list of commonly known metals, after gold, silver, and copper, is iron. Iron was discovered thousands of years ago and frequently used as a marker for the progress of human civilizations. This time of early discovery from pre-history until about 1100 BC became known as the Iron Age.

Iron, a reddish brown metal, is the "friendliest" of the metals. It readily combines with many other elements to form a huge variety of products known for their strength, since primitive times, like hand tools, cups, and plows to name just a few. Iron is the fourth most plentiful element in the Earth's crust making up about 5% of the elements present. Currently, more than 90% of all metal refined in the world is iron.

Iron is found in several ores, the main one being *hematite* (Fe_2O_3). Hematite has different colors and forms. The silvery black ore is used for jewelry and sometimes thought of as heavy, black "pearls." The red form of hematite is used in paint pigments and is known as the color, red ocher.

Alloys

Other metals, such as chromium, manganese, nickel, tungsten, cobalt, and chromium, combine with iron to make steel, an *alloy* of superior strength, hardness, and durability. About 99% of all iron mined in the world today goes into the manufacture of steel.

> **Alloys** are formed by the combination of two or more metals or a metal and non-metal.

When iron is coated with zinc, it is resistant to the rusting effects of oxygen and is said to be *galvanized.* Nails, wire, and large sheets of metal are treated this way to increase the life and use of such metals.

By mixing metals of different characteristics, in different proportions, a new metal or *metal alloy* can be created. The metals are melted together into a molten solution, cooled, and allowed to become solid again. The newly formed solid has the characteristics of its parent metals along with new properties, such as greater strength than either of its parents. (Example: steel = 80% iron, 12% chromium, 8% nickel; alloy manganese steel = 87% iron, 12% manganese, 1% carbon (railroad rails). Iron alloys are made by mixing together two or more molten metals and a non-metal. When producing 10-carat, 14-carat, and 18-carat gold, various percentages of gold, copper, and silver are used. **Table 12.2** gives examples of commonly used alloys and their composition.

Table 12.2 When metals are mixed in set quantities, alloys of specific properties are made.

Alloy	**Composition**
Brass	72% copper, 28% zinc
Bronze	93% copper, 7% tin
Carbon steel	1% manganese, 0.9% carbon, 98% iron
Manganese steel	12% manganese, 1% carbon, 87% iron
Stainless steel	12% chromium, 8% nickel, 80% iron
Chromium steel	3.5% chromium, 0.9% carbon, 95.6% iron
Sterling silver	92.5% silver, 7.5% copper
10 carat gold	42% gold, 12% silver, 40% copper
14 carat gold	58% gold, 24% silver, 17% copper, 1% zinc
18 carat gold	75% gold, 18% silver, 7% copper
24 carat gold	100% gold

Mercury is prized by scientists for its ability to dissolve other metals and form alloys. When combined with silver, dentists make a silver–mercury *amalgam* (alloy) to fill cavities in teeth. Mercury also dissolves gold and is used in the collection of gold from other ores.

Inner Transition Metals

The 30 elements of the inner transition elements are divided into two groups and called the actinide and lanthanide series.

The actinide series are difficult to isolate since they are highly unstable and experience radioactive decay. They also show a number of oxidation states. Uranium, for example, has compounds in each of the +3, +4, +5, and +6 oxidation states.

The lanthanide elements have electron configurations within the 4f and 5d energy levels. At these high energy levels, electron configuration is fairly theoretical. The lanthanides form 3+ ions as they go into an ionized state. It is assumed that ions are formed by losing the $6s^2$ and $5d^1$ (or 4f if 5d is not present) electrons. These metals are very similar to each other in their chemical properties. The lanthanide metals are shiny and are affected by water and acids.

Metalloids

Metalloids are elements that have the properties of both metals and nonmetals. Metalloids are also referred to as "semi-metals" and are located on the stair step borderline of the Periodic Table between metals and nonmetals. The metalloids are made up of boron, silicon, germanium, arsenic, antimony, tellurium, and polonium.

Silicon, the second most abundant element in the Earth's crust after oxygen, is found in granite, quartz, clay, and sand. At its most basic form, silicon has the same crystalline structure as a diamond. Maybe that is why when you walk along the beach and the sun is high, grains of sand seem to glisten like diamonds!

Silicon is used in semiconductors to carry an electrical charge. Silicon's natural conducting capacity is further increased when small amounts of arsenic or boron are added to it. Arsenic has one more valence electron than silicon and when it is mixed with silicon, this extra electron wanders around the crystal unattached and able to conduct electricity.

Figure 12.2 shows the location is metals and metalloids in the Periodic Table.

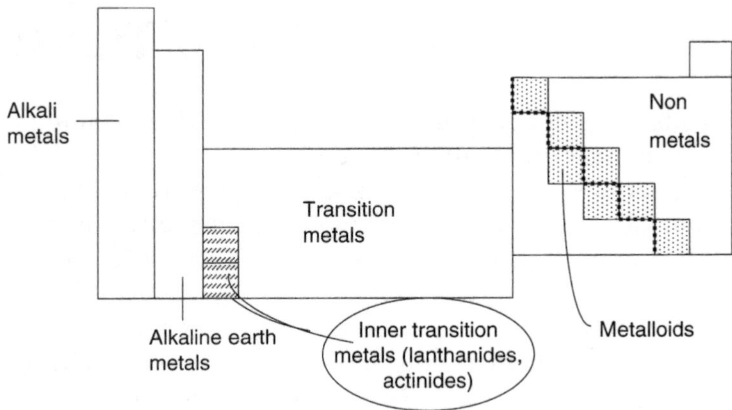

Fig. 12.2. Metalloids are found at the dividing line between metals and non-metals.

Non-metals

Put simply, elements that are not metals are non-metals. Non-metals are found at the far right-hand side of the Periodic Table and consist of only eighteen elements (counting from hydrogen with the non-metals to the right). Six of these elements are the noble gases in group VIII in the far right column. Non-metals are generally gases or solids that do not conduct heat or electricity well. Some non-metals, like diamonds (a stable crystalline form of carbon), are very hard and unreactive, while others, like chlorine, have free electrons and are more reactive. Non-metals are not lustrous and are not malleable or ductile. Eleven of the non-metals are gases, one is a liquid, and six are solids. **Figure 12.3** shows non-metals in solid, liquid, and gaseous forms.

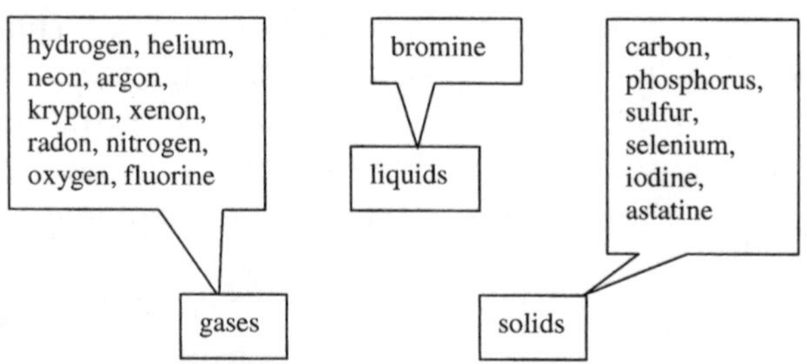

Fig. 12.3. Non-metals are found in all three forms of matter, liquid, solid, and gas.

Non-metals are described as elements without metallic properties. They are poor conductors of heat and electricity and are not magnetic like iron.

Noble gases are non-metals and exist as individual atoms such as neon and argon. Other non-metals are two-atom molecules, such as hydrogen, nitrogen, and oxygen.

Fluorine is at the top of the column (VIIA) in the Periodic Table known as the *halogens* or "salt formers." Chlorine, bromine, iodine, and astatine are also in the halogen group. Halogens easily accept electrons from other atoms and combine with metals to form salts. In nature, fluorine is found in ores of fluorspar (calcium fluoride) and cryolite (a combination of sodium, fluorine, and aluminum). In the body, fluorine is found in the blood, bones, and teeth. Many communities in the United States add small amounts of fluorine to drinking water, since the presence of fluorine has been found to prevent the formation of cavities in teeth.

Naming Metals and Non-metals

In general, metals form positive ions (cations), while non-metals form negative ions (anions). Noble gases, remember, are happy as they are and do not form cations or anions.

When a metal cation and a non-metal anion form a compound, they are called "metal–non-metal binary compounds." Metals with ions of only 1+, 2+, or 3+ charges in groups IA, IIA, and IIIA are called monoatomic, diatomic and polyatomic ions respectively. Lithium, potassium, and cesium all have a 1+ charge. Beryllium and strontium, in group IIA, are named with the same rules. However, when you name elements in groups IVA–VIIA, where anions of –1, –2, or –3 anions are formed, then "ide" is added to the name.

For example, in naming KCl, the metal is written first, potassium, followed by the non-metal, chloride; potassium chloride. To write the name of Co_2S_2, you first name the metal, which is cobalt, followed by the non-metal, sulfur, to get cobalt sulfide. See Chapter 9 for more naming examples.

POLYATOMIC IONS

You may wonder how to name a compound made up of two non-metals. The same rule applies. Write the non-metal that acts more like a metal, then the other non-metal. Usually, the quasi-metal (non-metal) is close to the zigzag border between metals and non-metals. Carbon is written first in carbon dioxide (CO_2). Hydrogen is always written first in compounds such as hydrogen fluoride (HF). Sulfur, nitrogen, and bromine are ordered in naming as in S_4N_4 or NBr_3. When several atoms are in a compound, Greek prefixes are used as given in Table 9.2. Examples of this are dinitrogen trioxide (N_2O_3) and carbon tetrafluoride (CF_4). The naming of other metals and non-metals in reactions will be described in later chapters.

Quiz 12

1. Which of the following element groups are most widely known?
   (a) halogens
   (b) metalloids
   (c) transition metals
   (d) lanthanides

2. Which element is the best conductor of electricity?
   (a) gold
   (b) silver
   (c) copper
   (d) aluminum

3. Most metals are naturally found
   (a) as pure metals
   (b) in limestone rock
   (c) in stream beds
   (d) combined with other elements in ores

4. Gold is
   (a) the most malleable and ductile metal
   (b) not used in jewelry
   (c) not an obsession for alchemists
   (d) always used to fill cavities in teeth

5. How do electrons behave in metal elements?
   (a) they are highly reactive
   (b) they form long, linear molecules
   (c) they float around in metal ions like broth
   (d) they react in ion pairs

6. What is the benefit of electron sharing?
   (a) the bonds are easily broken

(b) softer, more ductile alloys are formed
(c) inner shell electrons can be used
(d) it is a much more efficient use of energy

7. Which metal wins the "friendliest" award?
(a) sodium
(b) nickel
(c) mercury
(d) iron

8. Which two metals combine to give galvanized metal?
(a) lead and tin
(b) zinc and iron
(c) silver and aluminum
(d) lead and silver

9. Which of the following elements are stored in oil to prevent explosions?
(a) lithium
(b) actinium
(c) ytterbium
(d) strontium

10. When naming compounds
(a) metals come after metal salts
(b) two non-metals are named alphabetically
(c) metals come before hydrogen
(d) metals come before non-metals

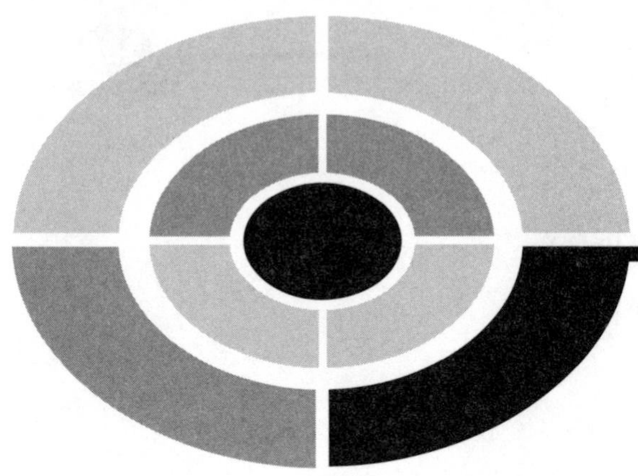

Test: Part Three

1. Atomic number (Z) provides
   (a) the number of protons in the nucleus of an atom
   (b) the weight of an element
   (c) a shortcut to make the study of chemistry easier
   (d) only the number of mesons in actinium
2. Covalent compounds are generally
   (a) brittle solids
   (b) lustrous hard solids
   (c) soft solids with high melting points
   (d) soft solids with low melting points
3. The lowest common multiple is easiest to use when
   (a) you have a graphing calculator
   (b) the charge from one ion is used as the multiplier for the other ion
   (c) your instructor helps you with the math
   (d) you have had a good breakfast first
4. In a binary covalent compound, boron is named before which of the following elements?
   (a) carbon
   (b) sulfur

(c) oxygen
(d) all of the above

5. In chemistry, the prefix *mono* is
(a) used to mean one
(b) generally not used for the first non-metal in the formula
(c) is a common illness found among college students
(d) used for multiplying sets of three

6. Sulfur hexafluoride is written as
(a) SF_3
(b) SH_2F
(c) SF_6
(d) SFO_2

7. Calcium, barium, and cadmium ions all have a
(a) +1 charge
(b) +2 charge
(c) +3 charge
(d) +4 charge

8. When a compound has the word thio as a prefix, it means it contains
(a) tin
(b) thallium
(c) sulfur
(d) selenium

9. Only electrons can
(a) bond with metals
(b) change to become ions
(c) carry a positive charge
(d) bond with noble gases

10. Copper forms how many kinds of ions with unique charges?
(a) 1
(b) 2
(c) 3
(d) 4

11. The largest group of carbon-containing compounds is called the
(a) halogen group
(b) organic group
(c) transition metal group
(d) inorganic group

12. Which of the following gives carbon at its simplest bonding form?
    (a) methane
    (b) ethane
    (c) propane
    (d) butane

13. When electron pairs are unequally shared between atoms of different elements, it is called
    (a) being greedy
    (b) sectional bonding
    (c) van der Waals forces
    (d) polarization energy

14. A compound is called a binary covalent compound when
    (a) only two elements are bonded together
    (b) more than three elements are bonded together
    (c) sixty carbons are bonded together
    (d) a dozen elements are bonded together

15. What does the prefix "bi" mean in bicarbonate?
    (a) two atoms of carbon are present
    (b) two atoms of calcium are present
    (c) hydrogen is present
    (d) oxygen is present

16. Carbon's six electrons fill the
    (a) 1s, 2s, and 2p orbitals
    (b) $1s^2$, $2s^2$, and $2p^4$ orbitals
    (c) $1s^2$, $2s^2$, and $2p^2$ orbitals
    (d) 1s, $2s^2$, and $2p^2$ orbitals

17. Monatomic ions can have
    (a) only fixed charges
    (b) fixed and variable charges
    (c) zero charge
    (d) only two charges per ion

18. Which of the following does not contain a double bond?
    (a) ethene
    (b) propane
    (c) polypropylene
    (d) butene

19. Sodium, potassium, and silver all have a
    (a) +1 charge
    (b) +2 charge
    (c) +3 charge
    (d) +4 charge

20. Dinitrogen trioxide is written as
    (a) NO
    (b) N_2O_3
    (c) N_2HO_3
    (d) N_2O_4

21. A double bond is normally shown with how many lines in a structural formula?
    (a) 1
    (b) 2
    (c) 3
    (d) 4

22. Petroleum crude oil comes primarily from the remains of
    (a) unprocessed mercury
    (b) iron filings
    (c) ice crystals
    (d) prehistoric plants and animals

23. To make ethanol from ethane, you must add which functional group?
    (a) $-CH_3$
    (b) $-H_2$
    (c) $-NH_2$
    (d) $-OH$

24. When a group of elements have many of the same characteristics and react the same, it is known as
    (a) hydrogen bonding
    (b) a homologous series
    (c) oxidation
    (d) reduction

25. The chemical formula for butane is
    (a) C_2H_2
    (b) C_6H_6
    (c) C_3H_{10}
    (d) C_4H_{10}

26. Who came up with the word isotopes for different forms of an element?
    (a) Hans Geiger
    (b) Lothar Meyer
    (c) Frederick Soddy
    (d) Marie Curie

27. Alpha (α) particles are
    (a) a neutral ion form
    (b) positively charged particles
    (c) negatively charged particles
    (d) a negative ion form

28. Magic numbers are the number of
    (a) the numbers to complete a perfect game of cards
    (b) elements currently in the Periodic Table
    (c) nuclear particles in a completed shell of protons or neutrons
    (d) protons found in every element

29. Radioactive uranium (^{238}U) decays eventually to
    (a) lead (^{82}Pb)
    (b) sodium (^{10}Na)
    (c) lithium (^{22}Li)
    (d) nitrogen (^{14}N)

30. Scintillation counters are used to
    (a) count coins in change machines
    (b) detect nuclear radiation from light flashes made by radiation hitting a sample
    (c) find wood floating in the water after a storm
    (d) count people going in to see a movie

31. Elements with atomic numbers greater than uranium (^{92}U) are called
    (a) noble gases
    (b) halogens
    (c) transuranium elements
    (d) alkaline bases

32. Radioactive elements are used in
    (a) toothpaste
    (b) ice cream
    (c) bicycles
    (d) submarines

33. "Heavy water" has a mass of
   (a) 12
   (b) 15
   (c) 20
   (d) 36

34. Non-metals are found where in the Periodic Table?
   (a) middle section
   (b) not in the Periodic Table
   (c) top left-hand corner
   (d) far right of the Periodic Table

35. When making 10-carat, 14-carat, and 18-carat gold
   (a) 15% nickel is added to pure gold
   (b) various percentages of gold, copper, and silver are used
   (c) 10% lead is added to make 18-carat gold heavier
   (d) silver is never used

36. When two or more metals or a metal and non-metal are combined it is called an
   (a) alloy
   (b) actinide
   (c) accident
   (d) aluminum isotope

37. Which solid metal can melt in your hand like candy?
   (a) molybdenum
   (b) cobalt
   (c) cesium
   (d) zirconium

38. What kind of fever affected thousands of Americans in the 1800s?
   (a) rhodium fever
   (b) arsenic fever
   (c) niobium fever
   (d) gold fever

39. Which ore is lead frequently found in?
   (a) bauxite
   (b) galena
   (c) rhodite
   (d) quartz

40. The years between 4,000 and 1,100 BC have been called the
    (a) Pre-Cambrian Period
    (b) Gold Rush
    (c) Bronze Age
    (d) Pre-MTV age

PART FOUR

Properties and Reactions

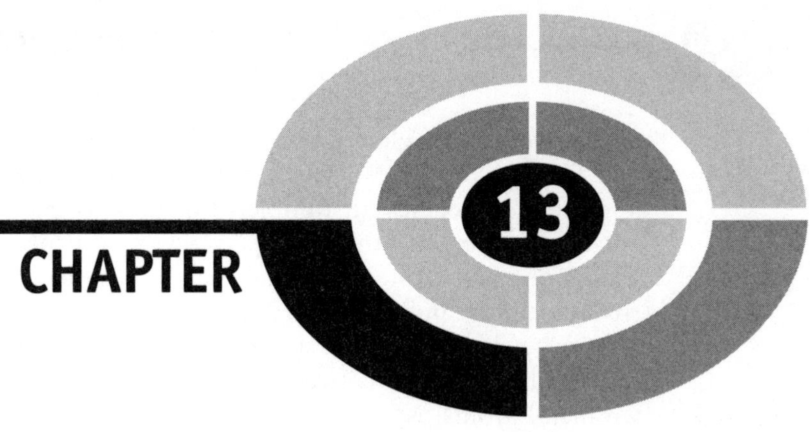

Chemical Bonding

Most elements are attracted to each other in a variety of ways. At the most basic level, protons, neutrons, and electrons are attracted in and around the nucleus by electrical attraction. Cutting-edge chemistry has discovered many subparticles smaller than these basic building blocks with additional attractive forces. A few hold outs, such as some of the noble gases, are unreactive in nature, but can be forced to share brief reactions.

> A **chemical bond** is the relationship between atoms within a molecule. The bonds are made through the interaction of electrons.

Ionic Bonding

One way to think of ionic bonds is to think of a direct gift; the total transfer of electrons between substances. When a metal forms an *ionic bond* with a non-metal, electrons are transferred from one element to another. When ions are formed, it is called *ionization*.

Ions are held by electrical attraction, sometimes called *electrostatic attraction*. This is like when socks stick together when they come out of the electric dryer. The molecules all tumbled together with the addition of heat create ionization. The same thing happens when you rub a rubber balloon on your head. Your hair proteins interact with the molecules of the rubber and electrostatic charge builds up. Try it!

As a general rule, ionic bonds occur between a metal and a non-metal. *Covalent* bonds occur between two non-metals. *Metallic* bonds form between two metal atoms.

EXAMPLE 13.1

A simple, ionic bond between a non-metal and a metal occurs in the formation of ordinary table salt, sodium chloride.

$$Na + Cl \Rightarrow Na^{+}\ \ Cl^{-} + \text{energy}$$

Oxidation of sodium happens when it loses electrons in the reaction, contributing to the ionization, and chlorine is reduced when it gains electrons in the reaction. As the transfer of the sodium electrons to the chlorine ion occurs, the ionic compound is formed. The positive (+) and negative (–) opposite charges attract and form an ionic bond.

Covalent Bonding

The sharing of an electron pair such as in *covalent bonding* is a great idea from an element's standpoint. It allows many more reactions to take place than would normally be possible. Remember, according to the octet rule the electrons in a shell must be filled by pairs to get a stable compound.

Table 13.1 compares and contrasts ionic and covalent bonding.

A covalent bond would not work without a contribution from one of the partners in a reaction. However, if a transfer of electrons does not occur, then sharing is an option.

Think of it as two college friends moving a heavy television that weighs 200 lbs. Friend A can lift 75 lbs easily, but friend B (who works out at the gym every day and is a collegiate discus thrower) can lift 125 lbs. Neither friend can lift and move the television by himself, but together, sharing the load and combining their strength, they can lift 200 lbs. If either drops their side (or severs the bond), it will be a big problem! So by sharing their strength and unique characteristics, something not possible individually is achieved. This is how covalent bonding works.

Table 13.1 Ionic and covalently bonded compounds have opposite strengths and weaknesses.

Ionic	**Covalent**
Metal–metal, non-metal–metal	Non-metal
Total transfer of electrons	Shared electrons
High melting and boiling points	Low melting and room temperature boiling points
Solids at room temperature	Liquids and gases at room temperature
Hard/brittle (inorganic compounds)	Relatively soft (organic compounds)
Strong bonds	Weak bonds
Electrically reactive	Does not normally conduct electricity
Soluble in water	Insoluble in water
Electron orbitals are separate	Electron orbitals overlap

The bonds between two hydrogen atoms with one electron each is called a *covalent* bond.

EXAMPLE 13.2
The covalent bond between hydrogen is simpler than ionic bonding

$$H + H \Rightarrow H : H \text{ or } H_2$$

The shared electron pair in a molecule is known as a *covalent bond* or a *covalent pair*. It is a friendly sharing union.

> **Covalent bonds** occur between atoms when they share electrons, but do not give them away totally. **Ionic bonds** occur between atoms when they give up electrons as a gift and they are not returned.

Non-polar Covalent Bonds

Covalent bonds between atoms of the same elements are known as *non-polar covalent* bonds. You can think of them as identical twin bonds. These bonds in the same family share electrons and are found in diatomic compounds such as H_2 and I_2. Non-polar bonds allow atoms to be more stable together than they are by themselves.

ELECTRONEGATIVITY

The ability of an atom in a covalent bond to pull shared electrons towards itself is called *electronegativity*. Group VII of the Periodic Table contains electronegative elements. The opposite of electronegativity is *electropositivity* or the ease with which an atom loses its shared electrons. Group I of the Periodic Table contains many electropositive elements.

Electronegativity = electron grabbing in a covalent bond.
Electropositivity = electron loss in a covalent bond.

Linus Pauling, an American chemist, was the first to take a closer look at the electrical difference of bonds. He tested the differences in energies of covalent bonds. In 1939, he published *The Nature of the Chemical Bond* which discussed the energy levels of molecules. Pauling was recognized for his work with protein structures. His electrochemical valency theory won the Nobel Prize for Chemistry in 1954.

While testing the electrical charge of molecules as they ionize and also combine into molecules, Pauling noted that elements all have different electronegativity values. In order to keep track of the different kinds of chemical bonds, he came up with a scale to record his results on different elements.

He found that fluorine was the most electronegative atom. He gave it an electronegativity number of 4.0. The least electronegative atoms of the alkali metals had values around 0.7 on Pauling's electronegativity scale. **Figure 13.1** shows the way electronegativity values increase compared to atomic number.

EXAMPLE 13.3

Look at the following example. Using the values in **Table 13.2**, can you determine the electronegativity values of potassium (K) and molybdenum (Mo)?

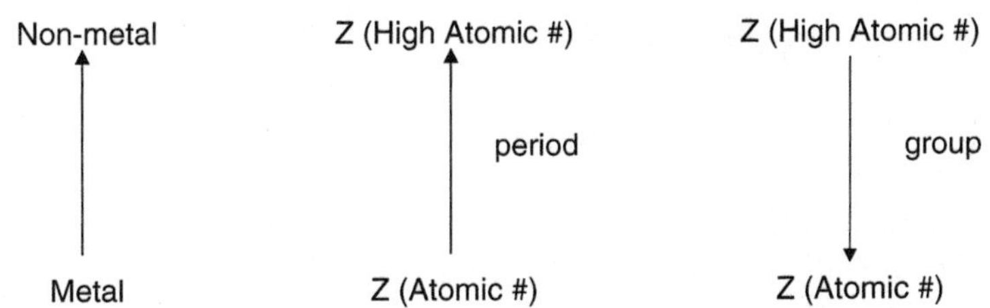

Fig. 13.1. Electronegativity values can be compared with atomic number.

Table 13.2 The Pauling scale of electronegativity helps to find an element's bonding potential.

Element	*Z*	Electronegativity value
Fr	87	0.7
Ca	20	1.0
Hf	72	1.3
Be	4	1.5
Co	27	1.8
Ge	32	1.8
Hg	80	1.9
Re	75	1.9
Sb	51	1.9
Te	52	2.1
At	85	2.2
Br	35	2.8
N	7	3.0
O	8	3.5

K, atomic number (Z) = 19

Mo, atomic number (Z) = 42

Which element has the higher electronegativity value? Did you get molybdenum? How about the electronegativity order of silver, oxygen, and beryllium from the highest value to the lowest?

Br, atomic number (Z) = 35; O, atomic number (Z) = 8; and Hg, atomic number (Z) = 80

Did you get oxygen, bromine, and mercury?

Polar Covalent Bonds

When the electrons are shared unequally between different elements, it is called *polar covalent bonding*. In a polar covalent bond, the bonding electrons spend more time near one atom than the other.

Just as the magnetic poles of the Earth pull ions to themselves more strongly than the equator, so polar bonded electrons are held more tightly by one atom than its bonding buddy. When this happens, a bond forms a positive (+) end with fewer electrons and a negative (–) end with more of the electrons in that area. This unequal sharing shape is called a *dipole*.

The three types of bonds can best be visualized as in **Figure 13.2.**

Dipole Moment

A *dipole moment* is the measurement of the charge separation between each part of a molecule. As polar bonds are unequally shared, they can be aligned in an electric field (like that between the poles of a magnet). The positive ends all line up with the negative side of the field and the negative molecule ends line up with the positive side of the field. When measurement of this activity is measured in a dipole moment and compared, chemists can guess at the geometry of a molecule's bonding and bond distance. By comparing the numbers, molecules with shared straight bonds (with lots of distance between the atoms), have a zero dipole moment with charges not being held more tightly, by one atom across the molecule, than the other.

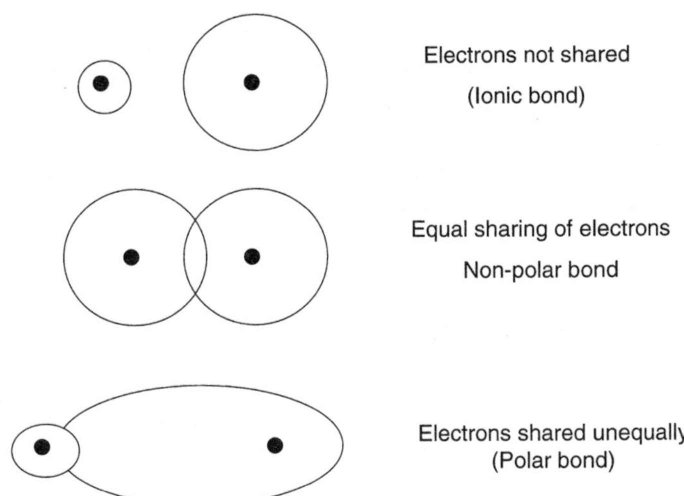

Fig. 13.2. Ionic, covalent, and polar covalent bond formation depends on several factors.

The distance the charges are separated from each other, the dipole moment, is measured in units of *debyes* (D). In a molecule like water, H_2O, the oxygen holds the electrons more tightly causing a polar separation and the dipolar moment is 1.94 D. In SI units, dipole moments are measured in coulomb meters (Cm),

$$1 \text{ D} = 3.34 \times 10^{30} \text{ Cm}$$

When molecules have an equal geometry, linear, equilateral triangle, or tetrahedron, they have zero dipole moment and are considered non-polar. This happens when all the atoms bonded to the central atom are identical, like carbon tetrachloride (CCl_4). When different atoms are bonded to the central atom, like $CHCl_3$, the molecule is polar.

Although calculating dipole moments is mostly used when doing advanced chemical equations, it is important to understand the kinds of tools chemists use to learn the details of element interactions.

Organic Molecules and Bonding

As we saw in Chapter 10, carbon–carbon bonds are strong and sturdy. Chains of long, carbon-based molecules form stable biological compounds. Carbon is the best example of the octet rule. With four valence electrons, it has a lot of choices. It may form four (1×) single bonds, 1 (2×) double bond and 2 (1×) single bonds, or 1 (3×), triple bond and 1 (1×) single bond. Notice in each of these molecules that the four valence electrons of carbon are happily bonded.

Figure 13.3 shows some examples of stable, unreactive carbon-based molecules with single, double, and triple bonds.

Single, Double, and Triple bonds

```
      H
      |
   H— C— H
      |
      H

     CH4
```

```
   H
    \
     C= O
    /
   H
          CH2O
```

```
H— C ≡ C — H
     C2H2
```

CH_4 CH_2O C_2H_2

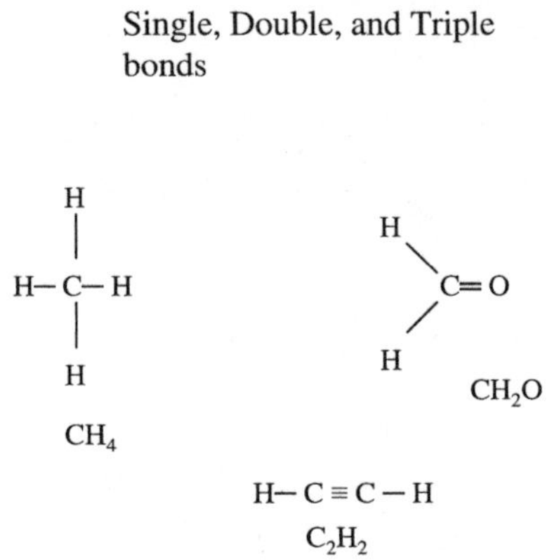

Fig. 13.3. Stable, carbon-based molecules can have single, double, and triple bonds.

Isomers

Berzelius who first used symbols to name elements, coined the word *isomer* through his work with chemical atomic weights. Not to be confused with isotopes, chemical element isomers have the same molecular formula, but different structural configurations.

> **Isomers** have the same molecular formulas, but different three-dimensional structures.

Think of yourself as a chief baker and molecules as your dough. The recipe for the dough is the same, the isomers are the different shapes the dough can take. It can be made into loaves, ropes, twists, rolls, or a dozen other shapes depending on what you want. The dough is the same, but the shape may cause it to react differently. If the dough is made into a long, thin rope, and cooked on a metal baking sheet that easily conducts heat, it will bake much faster than a round, solid core of dough baked in a thick-walled clay baker. If the same dough is used and both shapes are cooked at the same temperature for the same length of time, the thin rope will cook in half the time and probably burn, in the time it takes for the solid round loaf to bake. The shape makes a lot of difference in how the same formula reacts.

STRUCTURAL ISOMERS

There are basically two types of isomers, *structural isomers* and *stereoisomers*.

Isomers of pentane (C_5H_{12}) are shown in **Figure 13.4**. Though the groups are arranged differently structurally, the formula is the same. Chain isomers can be a few carbons long to many dozens of carbons in length.

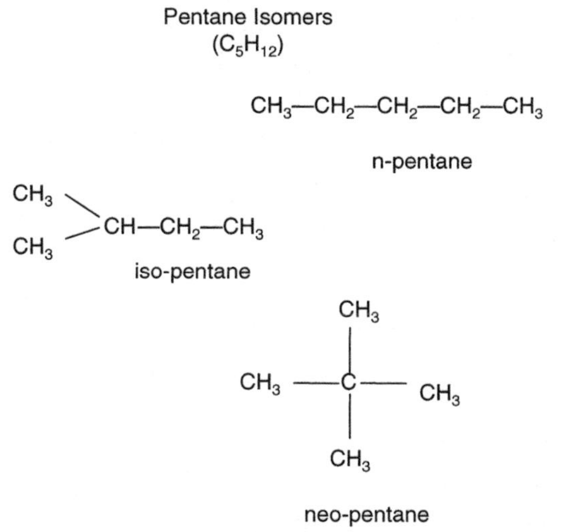

Pentane Isomers
(C_5H_{12})

$CH_3—CH_2—CH_2—CH_2—CH_3$

n-pentane

$(CH_3)_2>CH—CH_2—CH_3$

iso-pentane

$C(CH_3)_4$ (central C bonded to four CH_3 groups)

neo-pentane

Fig. 13.4. Pentane isomer groups are arranged differently, but the formula is the same.

STEREOISOMERS

Isomers that have the same formula, but their groups arranged differently in space are called *stereoisomers*. These groups can be located in many different places depending on the size and type of bonding. Stereoisomers are not structural isomers, they have their atoms combined in the same order, but are in different spacial positions. These isomers have the same functional groups, but again the placement is slightly different. **Figure 13.5** shows the

Stereoisomers

```
  H  H  H            H  H  H
  |  |  |            |  |  |
H-C--C--C-Cl       H-C--C--C-H
  |  |  |            |  |  |
  H  H  H            H  Cl H
```

1-chloropropane 2-chloropropane

Fig. 13.5. Stereoisomers of 1- and 2-chloropropane (C_3H_7Cl) have chlorine at different locations.

different placement of the chlorine group in stereoisomers of 1- and 2-chloropropane (C_3H_7Cl).

When ring structures like those found in benzene are considered, functional groups can be placed in several locations all around the ring. Remember, when compounds contain single and double bonds alternating around a ring structure, they are known as aromatic compounds as in **Figure 13.6**.

Ring Isomers

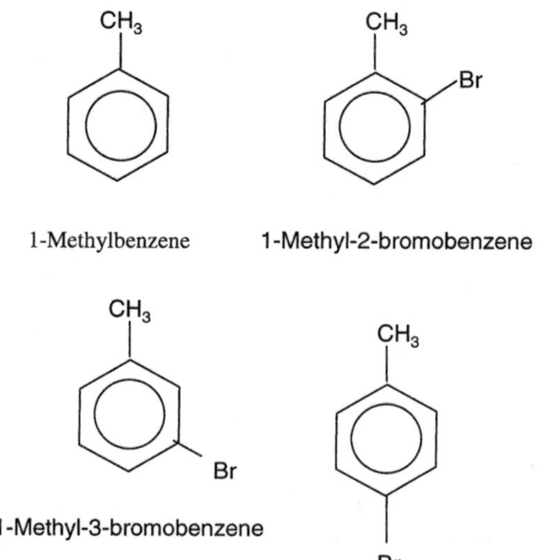

Fig. 13.6. Aromatic compounds vary according to the functional groups bonded around the rings.

gives examples of the different locations the functional groups can have around a ring structure.

CIS/TRANS ISOMERISM

Stereoisomers that have the same molecular formula, but different arrangements of their groups in space are called *cis/trans isomers*. These isomers are also sometimes called *geometric isomers*. The groups located on the same side of the molecule are in the *cis* formation, while the groups located on the opposite sides of the molecule put it into the *trans* formation. **Figure 13.7** shows the *cis/trans* structures of dichloroethene.

Cis/trans Isomers

Cl Cl
C=C
H H

Cl H
C=C
H Cl

cis –1,2-dichloroethene *trans* –1,2-dichloroethene

Fig. 13.7. *Cis/trans* bonding of dichloroethene shows the placement of chlorine atoms.

ENANTIOMERS

Isomers that are mirror images of each other are called *enantiomers* or *optical isomers*. Two *cis* isomers are mirror images of each other. When you hold up a mirror to one, it looks just like the other. You can put the image of the first *cis* isomer on top of the second *cis* isomer and they are identical or superimposable.

However, any two molecules that are mirror images of each other, but are not superimposable are called *chiral* molecules. The word *chiral* comes from the Greek word *cheir*, meaning "hand."

Take a pencil and outline your right hand. Now do the same thing with your left hand on another piece of paper. If you lay the right outline on top of the left outline, they will not be identical. Try it!

When the mirror image of a molecule is not identical to itself, this is thought of as "handedness" or a chiral formation. (Note: molecules or objects that *are* superimposable are called *achiral*.)

EXAMPLE 13.4
Which of the following are chiral (c)? Which are achiral (a)?

(1) fork, (2) ear, (3) hat, (4) spiral staircase, (5) shoe, (6) hammer.

Did you get: (1) a, (2) c, (3) a, (4) c, (5) c, (6) a?

Figure 13.8 shows the non-superimposable, chiral isomers of 2-butanol.

2-Butanol

$CH_3CHCH_2CH_3$ (OH on C-2)

```
     CH3                 CH3
      |                   |
OH — C — H           H — C — OH
      |                   |
     CH2                 CH2
      |                   |
     CH3                 CH3

            CH3
       CH3   |
        |    |
       OH — C — H
    H — C — OH
        |    |
        |   CH2
       CH2   |
        |   CH3
       CH3
```

Fig. 13.8. Non-superimposable, chiral isomers are like laying the right hand over the left hand.

The chemical bonding shown in this chapter is not all-encompassing, but should give you a general idea of the kinds of variations you will find in your study of chemical molecules.

Quiz 13

1. A chemical bond is a
   (a) good interaction between friends
   (b) poor possibility with alkali metals

(c) relationship between atoms in a molecule
(d) heat sensitive reaction inside the nucleus

2. Generally, ionic bonds form between
(a) two metals
(b) a metal and oxygen
(c) a metal and a halogen
(d) two non-metals and two carbons

3. Ionic bonding occurs
(a) with only one type of anion
(b) in the presence of high temperature
(c) when electrons are transferred from one element to another
(d) in the presence of uranium

4. Electronegativity describes the
(a) bad attitude of chemists when experiments don't work
(b) inability to maintain a charge
(c) electron loss in a covalent bond
(d) ability of an atom in a covalent bond to pull electrons to itself

5. The shared electron pair in a molecule is called
(a) ionic bonding
(b) covalent bonding
(c) non-polar ionic bonding
(d) isomer transfer

6. The ease with which electrons are lost in a covalent bond is thought of as
(a) electropositivity
(b) electronegativity
(c) valence exchange
(d) electromagnetism

7. Which American chemist first described the electrical difference of bonds?
(a) Antoine Lavoisier
(b) Stephen Hawking
(c) Albert Einstein
(d) Linus Pauling

8. Which of the following is the most electronegative element?
(a) gold
(b) silver

(c) bromine
(d) fluorine

9. A dipole moment
(a) describes the diatomic formation of two elements
(b) occurs when all electrons circle the nucleus equally
(c) is a measurement of the charge separation in parts of a molecule
(d) is composed of two polonium isotopes

10. When electrons are shared unequally
(a) no reaction occurs
(b) it is called polar covalent bonding
(c) the electrical charge is always positive
(d) only group IIA elements of the Periodic Table will react

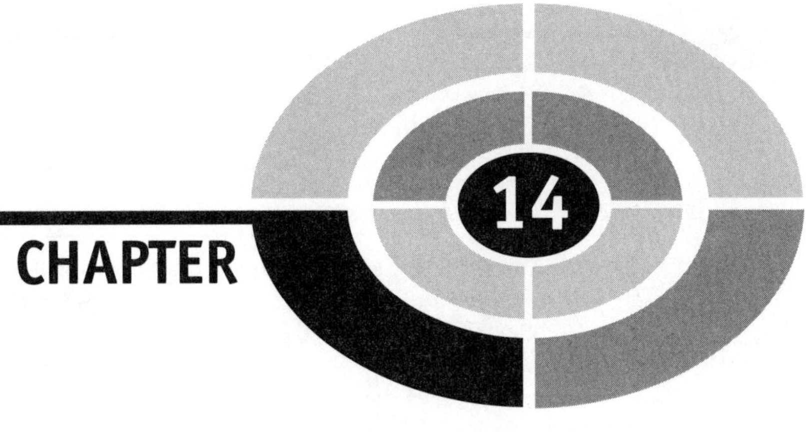

Acids and Bases

The most important thing to remember about acids is that they have very low pH values and burn skin, clothes, shoes, and almost everything they touch. They do not burn as a fire does with a flame, but by reacting strongly with the atoms of the substance and changing them permanently. Acids and bases cause chemical burns. Students are warned in their first laboratory safety class about strong acids and strong bases. Besides, (1) don't eat or drink anything in the lab; (2) don't run your fingers through the burner flame, and (3) don't blow yourselves up; always remember to (4) handle strong acids and bases *very carefully*.

Hydrochloric acid, HCl, is one of the first acids a beginning chemistry student learns. This strong acid reacts with most metals and forms hydrogen ions. Many experiments use HCl as a reactant. It is easy to spot the sloppy chemistry student, just look at his or her clothes and shoes. If they have a lot of tiny holes in their clothes, chances are good they don't have moths, but have splashed an acid or base on themselves while pouring. This is why chemists wear laboratory coats and eye goggles for protection. In this chapter, we will find out more about acids and bases and why they are so fierce.

Thousands of years ago, people learned that liquids were able to transform into different forms; grape juice fermented to wine and then if left too long to vinegar. In fact, the Greek word for vinegar is *acetum*, from which we get acetic acid, the main acid found in vinegar. *Acidus* is the Latin word for sour, a common characteristic of acids.

> An **acid** is any solution that releases hydrogen (H) ions when added to water and has a pH of less than 7.0. A **base** is any solution that releases hydroxide (OH) ions in water and has a pH of greater than 7.0.

pH

How do we figure out if a solution is acidic or basic? In 1909, a Danish chemist, P. L. Sorensen, calculated the negative logarithm of the hydrogen ion concentration. He wrote this as 1×10^{-n} with the pH equal to n. The name of the *pH scale* came from the French words for the "power of hydrogen" or "*pouvoir hydrogene*." The pH scale measures the amount of acidity in a solution.

> **pH scale** measures the acidity of a liquid by measuring the concentration of hydrogen ions. Neutral pH is equal to 7.0 on the pH scale.

The most often used piece of equipment in any laboratory beside the beaker or Bunsen burner is the pH meter. The pH meter has a special sensor that is inserted into a liquid sample to be tested. By passing a current through the sample and measuring the resistance and change in current, the amount of hydrogen ions (positively charged ions) can be determined. **Figure 14.1** shows a typical pH meter.

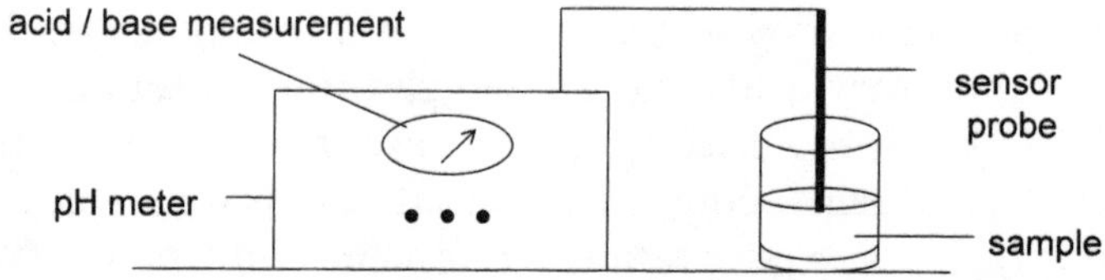

Fig. 14.1. A standard pH meter uses a probe to test the acidity of a sample.

Acids and Bases

Nearly 120 years ago, a Swedish graduate student in chemistry, Svante Arrhenius, came up with a few "rules of thumb" for deciding whether or not a solution is an acid:

1. In dilute solutions, acids taste sour. (*Never* taste a lab sample!)
2. Litmus paper changes from blue to red.
3. Acids react with metals like iron, magnesium, and zinc and release hydrogen gas.
4. When combined with bases, the products are water and salt.

To find out if a solution is basic, Arrhenius wrote that:

1. In dilute solutions, bases taste bitter. (But you are *never* going to taste a lab sample, right?!)
2. Litmus paper changes from red to blue.
3. Bases react with acids to form water and salt.
4. Bases feel soapy or slippery between the fingers and on the skin.

Acids

When an acid completely ionizes in water and gives up all its protons to water to form a hydronium ion, H_3O^+, it is considered a strong acid. **Table 14.1** lists some common acids from strongest to weakest.

Since acids are formed from the ionization of samples and the release of H^+ ions, acids that contain carbon held together by covalent bonds are generally much weaker.

Sulfuric acid is the most widely produced acid in the United States. It is used to make fertilizers (70%) from ores containing phosphate rock. The remaining 30% is used in processing industrial metals, oil refining, car batteries, and in general cleaners.

Boric acid is a weak acid used in eye washes and to clean glass. Boron is also a natural insecticide, since many insects cannot metabolize the boron molecule. Boric acid when mixed with equal parts white flour and powdered sugar makes a tempting insect bait for cockroaches. It causes internal blockage and death, while being safe for animals and children.

Table 14.1 Common acids from strongest to weakest should be handled with matching caution.

Name	Formula
Hydrochloric acid	HCl
Sulfuric acid	H_2SO_4
Nitric acid	HNO_3
Perchloric acid	$HClO_4$
Phosphoric acid	H_3PO_4
Acetic acid	CH_3COOH
Citric acid	$C_3H_5(COOH)_3$
Lactic acid	$CH_3CHOHCOOH$
Boric acid	H_3BO_3

Bases

Bases ionize almost completely in water. A base that does not ionize in water and is nearly insoluble in water may be thought of as a weak base, but it is really the amount of ionization that matters. $Ca(OH)_2$ is not particularly soluble, but the little that does dissolve is completely ionized, so it is called a strong base.

Sodium hydroxide (NaOH) is the strongest base that most chemistry students will use in the laboratory. The metal sodium (Na^+) forms a tight bond with the (OH^-) ion. It is used as a strong cleaner, to stabilize soil in road construction, and in the making of soap.

Magnesium hydroxide $Mg(OH)_2$ acts as a mild reactant by gently neutralizing stomach acids. When spicy or acidic foods, like tomatoes, cause problems, an antacid containing magnesium hydroxide neutralizes the acid and brings speedy relief. **Table 14.2** lists a few common bases.

Table 14.2 It is just as important to become familiar with chemical bases as with acids.

Name	Formula
Sodium hydroxide	$NaOH$
Potassium hydroxide	KOH
Calcium hydroxide	$CaOH$
Magnesium hydroxide	$Mg(OH)_2$
Calcium carbonate	$CaCO^3$
Ammonia	NH^3
Calcium diphosphate	$Ca(H_2PO_4)_2$
Barium sulfate	$BaSO_4$

Brönsted–Lowry Acids and Bases

In 1923, two chemists, Johannes Brönsted and Thomas Lowry, described acids and bases in the scientific literature. They were studying how the transfer of hydrogen ions (protons) took place between reacting molecules. A Brönsted–Lowry acid donates a proton in a reaction, while a base is on the receiving end of the proton transfer. In the Brönsted–Lowry definition of acids and bases, ions as well as larger more complex molecules are included. The pairs of (H_2O, OH^-) and (NH_3, NH_4^+) are called Brönsted–Lowry *conjugate acid–base pairs*.

EXAMPLE 14.1

A Brönsted–Lowry reaction is shown. Which is the acid? Which is the base?

$$\underset{\text{base}}{NH_3} + \underset{\text{acid}}{H_2O} \rightarrow \underset{\text{acid}}{NH_4^+} + \underset{\text{base}}{OH^-}$$

When an acid reacts and loses a hydrogen ion, it forms a base. In the above reaction, the H_2O acts like an acid, loses a hydrogen ion (H^+) and becomes a base (OH^-). The base, NH_3, gains a hydrogen ion (H^+) and becomes more acidic (NH_4^+).

> A **conjugate base** of a sample or ion occurs when an H^+ ion is lost. A **conjugate base** of a sample or ion occurs when an H^+ ion is added.

Some acids and bases are chameleons and can accept protons in one reaction and then turn around and donate them in a subsequent reaction. A lot of physiological reactions work like this with water or blood serving as the go-between solution. Ions or molecules, which can swing either way, depending on the environment and company present, are called *amphiprotic*. They can either lose or add a proton (H^+) in a reaction.

An ion or molecule is *amphoteric* when it can serve as either an acid or a base in a reaction, but has no protons (H^+).

To bring the Brönsted–Lowry acid and base idea all together, remember the following:

1. A base accepts protons.
2. An acid provides protons in a reaction.
3. Acidic and basic reactions don't just occur with protons (H^+).
4. Ions as well as molecules can be acidic or basic.
5. Some reactants can swing either way, providing or accepting protons.

As shown in **Table 14.3**, acids and bases have different strengths and weaknesses. Some mild acids like water (remember, it provides a proton) can be used to wash your hands of dirt. You couldn't use sulfuric acid for the same purpose.

The strengths of acids can be found by how well they ionize. If two acids react with a certain solution, generally one will be ionized more than the other. Just as people are all individuals, elements have strengths and weaknesses too. Reactions usually go in the direction of the weaker acid or weaker base. A stronger acid will be transformed into products that include a weaker acid. The same is usually true of reactions with bases.

One exception happens in the presence of water. Water bends the rules. In water, strong acids seem to have the same strengths. They ionize well and seem to "even up" the differences.

You may see the words *conjugate acids* and *conjugate bases*. A *conjugate acid* is the part of an acid–base reaction that donates the proton. A *conjugate base* is the part of the joined compound that can accept a proton.

Table 14.3 Many foods and household solutions are strongly acidic or basic.

Name	pH
Hydrochloric acid	2.0
Stomach acid	1.0–3.0
Lemon juice	2.2–2.4
Vinegar	2.4–3.4
Carbonated drinks	3.9
Beer	4.0–4.5
Milk	6.4
Blood	7.4
Sea water	7.0–8.3
Baking soda	8.4
Antacid	10.5
Cleaning ammonia	11.9
Bleach (sodium hydroxide)	14.0

Strong acids have weak conjugate bases. **Strong bases** have the weakest conjugate acids.

The strength of acids and bases is based on the amount of hydronium ions (H_3O^+) and their concentration in a solution. The pH meter records this concentration with strong acids ranging from 2.0 to 5.0, neutral pH being a pH of 7.0, and strong bases ranging in values from 8.5 to 10.0. The lower the pH number, the stronger the acid; the higher the pH number, the stronger the base.

There are two clues to figuring out the strength of an acid:

1. Check the polarity of the H^+ bond. The more polar the bond, the more easily the H^+ proton is removed and the stronger, more interactive the acid.
2. Check the size of the atom bonded to the H^+.

Commonly, you will find that the larger the atom, the weaker the bond and the stronger the acid. When hydrogen bonds to chlorine, or bromine, the larger molecule with greater orbitals has more interactivity. When comparing the acids HF, HCl, HBr, and HI, the acidity increases as the size of the atoms increases. So they can be arranged as $HI > HBr > HCl > HF > H_2O$.

The strength of bases can be found in much the same way. Stronger bases ionize almost completely in water, while weak bases do not. This happens because less ionized bases do not have free orbitals to allow the acceptance of additional protons. (Everybody is comfortable and stable and resists change.) A stronger base is able to accept H^+ more easily than a weaker one because it goes into the ion state more easily. The following bases can be arranged by strength $OH > NH_3 > HCO_3 > C_2H_3O_2 > NO_3 > HSO_4$.

Neutralization

As anyone who has ever gotten a strong acid on the skin can tell you, the reaction with skin molecules is universally a bad thing. But let's think about it. If an accident happened and you or your lab partner were splashed with acid, would you know how to stop the reaction?

In a laboratory, get under water as quickly as possible. Whether to wash your hands, use the eye wash, or be doused in a full body shower, the action of the water dilutes the acid. This has a *neutralizing* effect on the reaction between the acid/base and the skin or the clothes. Another option would be to use a base to counteract the acid, but this usually takes too much time.

> Canceling out of acids by bases and bases by acids is called **neutralization**.

Water is the best response for strong bases, too. Since they are fully ionized, they can cause the same kinds of chemical burns as acids. Sometimes students forget this and treat bases with less respect than acids. This is not a good plan. Strong acids cause the same kinds of bad burns as strong bases. Again, if in contact with a base, another option would be to counteract the base with an acid.

The key is that acids cancel out the affects of bases and bases cancel the affects of acids.

Oxidation Numbers

As we learned earlier, electrons are often shared in compounds. Sometimes they are shared fairly and sometimes one or the other of the atoms will pull electrons tighter and have more control of the electrons. Chemists keep track of this control of electrons by something called *oxidation state* or *oxidation numbers*. When atoms are surrounded only by atoms of their own kind, they share equally, but it depends on the strength of the bonding of the different atoms. In general, if control is strong and is increased in a reaction, the sign is negative. If electron control is weaker, the sign is positive.

> **Oxidation numbers** are based on the difference between the number of electrons an atom in the element can control and the number of electrons an atom in a compound can grab and hang on to.

Table 14.4 gives tips to help you find the oxidation numbers of elements and compounds.

Buffers

A very important part of chemistry is learning to use buffers. A buffer gives you "wiggle room" in the pH of a solution. Buffers are usually a pair of compounds like HCO_3^- and H_2CO_3 that keep the pH nearly constant from the starting pH. Buffers are important since they minimize huge swings in pH that occur with the addition of acids or bases. A buffer solution usually contains either a weak acid and its salt or a weak base and its salt, which is resistant to changes in pH.

Think of it like going to the market. Your first few packages are easy to hold. As you continue to buy more and more, you shuffle and re-shuffle the bags, boxes, and envelopes. You can maintain balance and hold all your

Table 14.4 Here are some simple ways to figure out oxidation numbers.

Ways to figure out oxidation numbers ***(but were afraid to ask)***
1. Elements that are bonded to themselves have an oxidation number of zero (Br_2). 2. Group IA (alkali metals like lithium) have oxidation numbers of +1. 3. The oxidation number of hydrogen is usually +1 (except in hydrides, where hydrogen ions bond to metal ions and have an oxidation number of −1). 4. Group IIA (alkaline earth like calcium or barium) have oxidation numbers of +2. 5. One element that never changes oxidation number is fluorine. It has an oxidation number of −1. 6. Most often, oxygen has an oxidation number of −2 (except in peroxides like H_2O_2, then oxygen has an oxidation number of −1). 7. Halogens, like chlorine, bromine, and iodine have oxidation numbers of −1 (like fluorine), but only when bonded to elements of weaker electronegativity. 8. Elements of Group IIIA have oxidation numbers of +3. 9. Elements of Group IVA can have oxidation numbers of +2, +4, and −4. 10. Elements of Group VA (like nitrogen) can have a wide range of oxidation numbers of +1 to +5, 0, and −1 to −3 (NH_3) depending on the bonding. 11. Elements that bond to elements of Group VIA (like sulfur) have oxidation numbers of −2, 0, +4, and +6 (H_2SO_4).

purchases, to a certain point, but when you reach your limit, the whole thing tips and falls to the ground.

> **Buffers** are compounds used to react with hydrogen ions (H^+) and hydroxide ions (OH^-) in order to have a "toning down" or neutralizing effect.

Similarly, buffer capacity can't go on forever. Just as you can't carry an infinite number of purchases, so buffers will eventually reach capacity. The solution suddenly goes from being balanced at a certain pH, to being very acidic or basic depending on whether an acid or base was steadily added.

Most of chemistry focuses on the effects of acids and bases on elements' makeup. When you are familiar with the way acids and bases behave, you'll have a good handle on understanding chemistry.

Quiz 14

1. Which of the following is not a property of an acid?
   (a) has a pH >7.0
   (b) releases hydrogen (H^+) ions when added to water
   (c) causes chemical burns
   (d) none of the above

2. A base
   (a) has a pH <7.0
   (b) does not cause chemical burns
   (c) does not ionize in water
   (d) is any solution that releases hydroxide (OH^-) ions in water

3. The pH scale measures
   (a) the weight of a liquid
   (b) the acidity of a liquid
   (c) the temperature of a liquid
   (d) the density of a liquid

4. Which of the following is not a "rule of thumb" for deciding whether or not a solution is an acid?
   (a) in dilute solutions, acids taste bitter
   (b) litmus paper changes from blue to red
   (c) acids react with metals like iron, magnesium, and zinc and release hydrogen gas
   (d) when combined with bases, the products are water and salt

5. Which of the following is not a "rule of thumb" for deciding whether or not a solution is a base
   (a) in dilute solutions, bases taste bitter
   (b) litmus paper changes from red to blue
   (c) bases react with acids to form water and salt
   (d) bases feel sticky and tacky between the fingers and on the skin

6. An acid is considered a strong acid when
   (a) combined with a base and gives off a very pungent odor
   (b) it has a pH value >9.0
   (c) it completely ionizes in water and gives up a proton to water to form a hydronium ion, H_3O^+
   (d) it has strong conjugate base

7. A Brönsted–Lowry acid
   (a) changes litmus paper from red to blue
   (b) donates a proton in a reaction while a base is on the receiving end of the proton transfer
   (c) is a classification of a strong acid
   (d) ionizes almost completely in water

8. To be amphoteric
   (a) a stronger acid will be transformed into products that include a weaker acid
   (b) it is the element in the reaction that oxidizes another element while at the same time being reduced itself
   (c) a stronger base ionizes almost completely in water, while weak bases do not
   (d) an ion or molecule can serve as either an acid or base in a reaction, but has no protons (H^+)

9. Oxidation is
   (a) when a compound loses oxygen, gains hydrogen or gains electrons
   (b) when a solution releases hydrogen (H^+) ions when added to water
   (c) when a compound gains oxygen, loses hydrogen, or loses electrons
   (d) when acids and bases can accept protons in one reaction then turn around and donate in a subsequent reaction

10. Chemists Johannes Brönsted and Thomas Lowry
    (a) created litmus paper
    (b) discovered "buffers" as sets of compounds that react with and occupy hydrogen ions (H^+) and hydroxide ions (OH^-)
    (c) established the pH scale
    (d) described acids and bases while studying how proton transfer occurs

CHAPTER 15

Solids

A brick, a leaf, and a knee cap, what do these three things have in common? Can you guess? They are all solids. They have different functions and a much different make-up, but they are all members of the solid club.

Solids are, well, solid. They are made to last. When archeologists excavate ancient sites, solids are still there, but liquids and gases are long gone. We base 90% of what we know of past civilizations on the solids they left behind like bits of pottery, metal tools, and weapons. But why are some types of matter members of the solid club and not others?

We saw back in Chapter 3 that all matter comes in one of three types, solid, liquid, or gaseous. Figure 3.2 shows how the main difference between solids and other forms of matter is density. The more compact or dense the atoms or molecules of a sample, the more tightly packed and the more solid it is.

Solids are denser than liquids or gases, but there is a lot more to it than that. Solids have a lot of quirks that make the club fairly interesting.

Density

Compared to liquids, the atoms of a solid are a lot more compressed. The atoms that make up a solid are bonded very tightly and have very little room to move around. They are like people on a packed commuter train at rush hour on a Monday morning. They are basically stuck in one spot until some outside force allows them to move more freely. These outside forces are commonly, changes in bonding, crystallization, heat, and pressure.

Depending on their location in the Periodic Table, different elements can have very different densities. For example, the density of balsa wood is approximately 0.13 g/cm^3 and the density of water (20°C) is 0.998 g/cm^3. It makes sense then that porous balsa wood floats easily in water since its density is so much less. The density of copper is 8.96 g/cm^3, compared to the density of mercury at 13.55 g/cm^3. If a copper penny is dropped into a container of mercury, would it sink or float? Well, look at the densities. Mercury is much denser, so the copper penny would float. If the same penny were dropped into a beaker of water, what would happen then?

Amorphous Solids

Solids are normally found in one of two types, amorphous and crystalline. The first group, amorphous, is made up of shifting members that really can't make up their minds if they like being a solid or not. These are things like wax, rubber, glass, and polyethylene plastic.

> **Amorphous solids** have no specific form or standard internal structure.

Brittle, non-crystalline solids tend to shatter every which way into sharp pieces when broken. These amorphous solids are less dense than other crystalline club members and have no definite melting point. When heated, they slowly soften and become very flexible. Have you ever heated glass tubing in a flame to make a 90° angle in a straight rod? This is an example of the shape changing of non-crystalline solids.

Crystalline Solids

Most people are a lot more familiar with crystalline solids. These are solids like quartz, diamond, salt, and different gemstones. The atoms of crystalline solids go together into specific crystal patterns of an ordered lattice or framework.

Crystalline solids are arranged into regular shapes based on a cube; simple, central, and face-centered.

Figure 15.1 shows the three ways the atoms of a crystalline solid can be arranged. As a molecule goes from a simple cubic structure to a face-centered cubic structure, the density increases. The less space between the atoms, the more tightly packed the entire molecule, and the harder and less flexible. Unlike amorphous solids, a lattice structure provides for predictable breaks along set lines. This is the reason why diamonds and gemstones can be cut into facets. The round, oval, pear, emerald cut, and diamond-shaped cuts used in jewelry can be cut by different gem cutters all over the world due to their characteristic lattice structures.

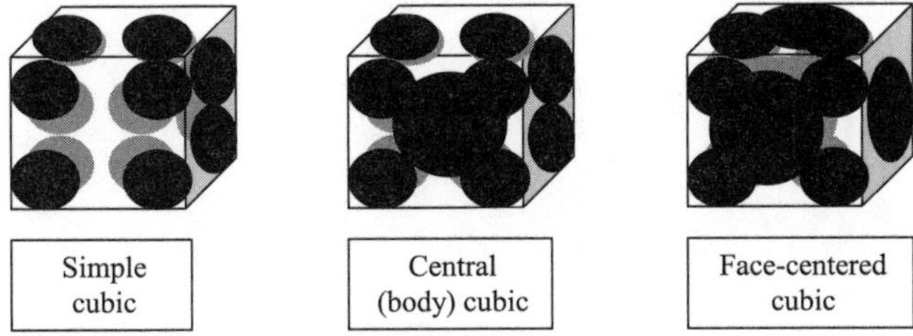

Fig. 15.1. The beauty of crystalline solids comes mainly from their atomic arrangements.

Crystallization and Bonding

There are four different types of bonding that occur in crystalline solids. These determine what type of solid it is.

The four types of solids are **molecular**, **metallic**, **ionic**, and **covalent**.

Molecular Solids

These types of crystalline solids have molecules at the corners of the lattice instead of individual ions. They are softer, less reactive, have weaker non-polar ion attractions, and lower melting points.

A *molecular* solid is held together by *intermolecular* forces. These intermolecular forces were described in Chapter 13. The bonding of hydrogen and oxygen in frozen water shows how hydrogen forms bonds between different water molecules.

Figure 15.2 shows the cubic arrangement of sodium chloride.

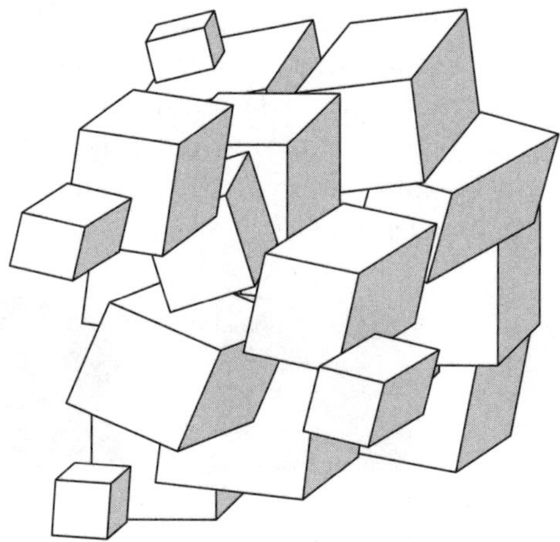

Fig. 15.2. The cubic arrangement of sodium chloride gives an example of intermolecular forces.

METALLIC SOLIDS

Another type of crystalline solid is made up of metals. All metals, except mercury, are solid at room temperature. The temperature needed to break the bonds between positive metal ions in specific lattice positions, like iron in iron(II) disulfide (FeS_2), and the happy valence electrons around them is fairly high. This strong bonding gives their stable molecules flexibility and allows them to be formed into sheets and strands without breaking as we learned in Chapter 12.

A *metallic* solid like iron or silver is held together by the "bread pudding" type of bonding. When a positive central core of atoms is held together by a surrounding general pool of negatively charged electrons it is called *metallic bonding*. This arrangement of (+) metal ions and delocalized valence electrons makes them good conductors of electricity.

IONIC SOLIDS

Ionic solids form a lattice with the outside points made up of ions instead of larger molecules. These are the "opposites attract" solids. The contrasting forces give these hard, ionic solids (like magnetite and malachite) high melting points and cause them to be brittle. **Figure 15.3** shows the melting points of different solids.

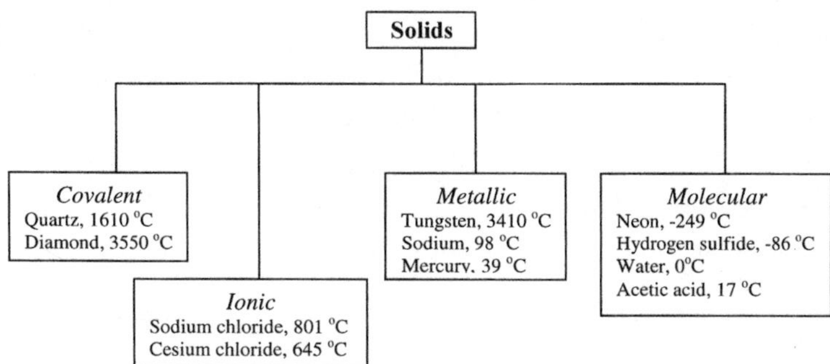

Fig. 15.3. Depending on bonding strength, solids have a wide variety of melting points.

Ionic bonding in a solid occurs when anions (–) and cations (+) are held together by the electrical pull of opposite charges. This electrical magnetism is found in a lot of salts like potassium choride (KCl), calcium chloride (CaCl), and zinc sulfide (ZnS).

Ionic crystals that contain ions of two or more elements form three-dimensional crystal structures held together by strong ionic bonds.

Additionally, ionic compounds are *electrolytes*. Electrolytes form ions and carry a current when melted or dissolved in a solvent during *electrolysis*.

> **Electrolysis** is the method of breaking down an electrolyte by separating its ions between positive and negatively charged electrodes.

Positive cations move to the cathode (+) and gain electrons, while negative anions move toward the anode (–) and lose electrons.

COVALENT SOLIDS

A grouping of covalent bonds holds some solids together. Assembled together in large nets or chains, covalent multi-layered solids are extremely hard and stable in this type of configuration. For example, diamond atoms use this type of structure when they arrange into three-dimensional solids. One carbon atom is covalently bonded to four other carbons. This strong crystalline structure makes diamond the hardest known organic solid.

Covalent crystals are all held together by single covalent bonds. This type of stable bonding produces high melting and boiling points.

The different bonding and forms of carbon in a diamond (pyramid shaped), graphite (flat layered sheets), or buckminsterfullerene (C_{60} and C_{70}, shaped like a soccer ball) give an idea of the variety and stability of covalent molecules. Nets, chains, and "balls" of carbon bonded into stable molecules make these solids hard and stable.

> **Allotropes** are different structural forms of the same element. Graphite, diamond, and buckminsterfullerene are all allotropes of carbon.

Physical Characteristics

Solids have regular parts of their personality that can be used to classify them. These can be found by their placement in the Periodic Table as well. The metals found in the middle of the Periodic Table share a lot of the same characteristics, like brothers and sisters in the same family that have the same hair or eye color. Some of these characteristics are hardness and structure (crystalline or amorphous), melting point, and electrical conductivity. **Table 15.1** lists these personality traits of solids.

Table 15.1 The individuality of solids can be found in their variety of characteristics.

Solids
Fixed volume Definite structure Cannot be compressed Held together by molecular, ionic, metallic, or covalent bonds Do not expand when heated Amorphous or crystalline in form High density

Temperature

One of the major players in the solid club is temperature. Temperature controls a lot of who gets along with whom and whether they play nice. Temperature also decides which form club members take to be a member.

Look at water, normally a member of the liquid club, but exposed to very low temperatures and pow! Ice. Not only a solid, but see through as well. Thanks to this special gift of H_2O, we enjoy popsicles, ice hockey, and snow balls.

What happens when we raise the temperature a few degrees? No more ice. It's back to the liquid club. Without ice, we have skiing on rocks (which is very hard on skis), no polar ice caps, and soupy chocolate chip ice cream. In other words, the end to life as we know it on this planet!

Generally, every element can be a solid if the temperature is dropped low enough, even gases. This helps us to understand the conditions on planets like Jupiter and Saturn in our solar system. They are mostly made up of frozen gases.

In the next two chapters, we will see the differences between solids, liquids, and gases. Differences are good! They make the study of chemistry fun.

Quiz 15

1. Amorphous solids
   (a) are generally very dense
   (b) have no specific form
   (c) become very rigid when heated
   (d) melt very fast when heated

2. Which of the following is not true of a crystalline solid?
   (a) they are arranged into regular shapes based on a cube; simple, central, and faced centered
   (b) they can be predictably cut or broken along set lines
   (c) they tend to be very unstable and can change state very easily
   (d) the atoms go together into specific crystal patterns of an ordered lattice or framework

3. Which of the following is not a type of solid?
   (a) anatomic
   (b) metallic

(c) covalent
(d) ionic

4. Stable molecules that give them flexibility and allow them to be formed into sheets and strands without breaking is a property of what type of solid?
   (a) anatomic
   (b) metallic
   (c) covalent
   (d) ionic

5. A solid that forms a lattice with the outside points made up of ions instead of large molecules is what type of solid?
   (a) anatomic
   (b) molecular
   (c) covalent
   (d) ionic

6. Which is not a property of a covalent solid?
   (a) the very stable bonding produces high melting and boiling points
   (b) the contrasting forces give these solids high melting points
   (c) they are held together by single covalent bonds
   (d) nets, chains, and "balls" of carbon bonded into stable molecules make these solids hard and stable

7. The role of temperature in a solid is
   (a) secondary to pressure
   (b) only important to transition metals
   (c) a major player in what form an element takes at room temperature
   (d) insignificant

8. Metals found in the middle of the Periodic Table
   (a) share a lot of the same characteristics
   (b) rarely react with other elements
   (c) have almost all the same properties of gold
   (d) are limited to reactions with oxygen

9. Gases are solids when
   (a) there is an excess of oxygen in the reaction
   (b) combined with lead or zinc
   (c) not in the liquid phase
   (d) the pressure is high and the temperature lowered to sub-zero levels

10. Molecules in solids
    (a) are always shaped into a lattice
    (b) have very little movement due to high density
    (c) bond easily to carbon
    (d) are seldom organic in composition

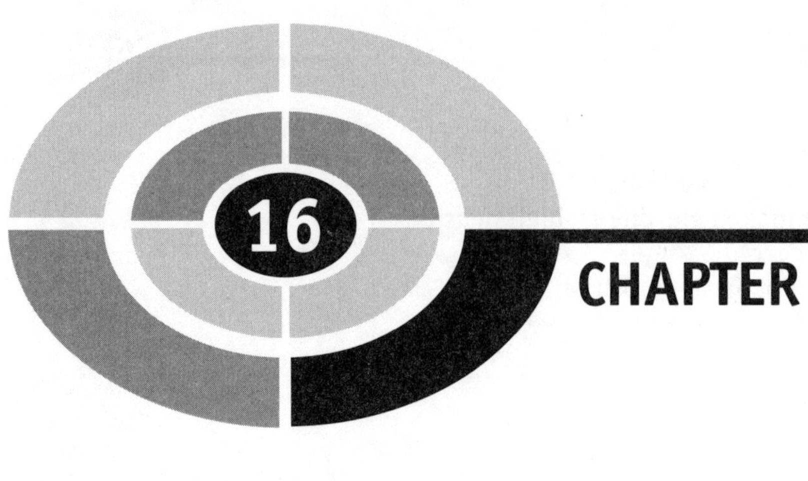

Liquids

Lava, a cold mountain stream, and mercury are all liquids. They are very different, with different compositions and different standard temperatures, but they are all members of the in-between element club, known as the liquids. The members of this club are affected by environment much more than solids. Sometimes the difference of a few degrees of temperature can cause an element to slip from being a member of the solid club to new membership in the liquid club. Cesium has a particularly great trick of switching from a solid to a liquid. It is solid at room temperature, but when held in the hand and heated to body temperature of 98.6°F, it melts and becomes a liquid.

A chocolate bar does the same thing. On the shelf at room temperature, it's a solid, but when held in the hand and heated to body temperature, it melts.

Table 16.1 gives some general characteristics of liquids.

Density

As seen in Figure 3.2, liquids are less dense than tightly packed solids. The density of a liquid affects the way a liquid acts. It has different characteristics from solids, like the ability to flow across a table top when spilled.

Table 16.1 Liquids are the stars of the in-between club of matter.

Liquids
Fixed volume
Loose structure
Cannot be tightly compressed
Have different viscosities or resistances to flow
Held together by surface tension
Miscible or immiscible
Expand and vaporize when heated
Unequal molecular bonding
Medium density

The density of water is 1.00 g/ml at 4°C. The metric system of measuring liquid density is based on this number. When comparing the density of liquids, generally they can be compared to water. This makes it easier to figure out whether liquids will mix or not, since two liquids of very different densities don't usually combine.

There are exceptions. Very dense ionic solutions like salt water will dissolve in water since both are polar. Oil which is non-polar will not dissolve in water even if the densities were close to each other. Their failure to mix is due to their properties, rather than density.

For example, the densities of mercury (13.5 g/ml) and water (1.0 g/ml) are very different relative to each other. This *relative density* difference (sometimes called *specific gravity*) causes mercury to sink to the bottom of a container filled with water.

> **Relative density** (specific gravity) is the ratio of the density of a sample at 20°C divided by the density of water at 4°C.

Some of the physical characteristics that affect liquids are viscosity, surface tension, boiling point, vaporization, condensation, and evaporation. These are described below.

Viscosity

The shape and combination of ions and molecules in solids determines a lot about their characteristics. The same is true of liquids. The size, strength, and shape of molecules along with the intermolecular forces in liquids have a big effect on the *viscosity* of different liquids.

Think of how ketchup pours when we are hungry for French fries. Ketchup is more viscous than water. Water and wine pour easily because they have a low viscosity.

EXAMPLE 16.1
Look at the following liquids. Can you name which have high viscosity (hv) and which have low viscosity (lv)?

(1) pancake syrup, (2) vinegar, (3) motor oil, (4) apple juice, (5) molasses, and (6) pine sap.

Did you get (1) hv, (2) lv, (3) hv, (4) lv, (5) hv, and (6) hv?

> **Viscosity** is the capability of a liquid to flow or not flow freely at room temperature.

Think of the difference between water and honey. Hydrogen and oxygen contain covalent bonding, but there are intermolecular forces of attraction between water molecules called hydrogen bonding. Commonly, these are stronger than the forces between organic molecules, like proteins and the sugars in honey. However, if the molecules are big enough, there can be very strong intermolecular forces between organic molecules. This is what happens with honey, giving it a very high viscosity. The stronger the molecular forces between molecules, the thicker (or more like a solid) a liquid becomes. The weaker the molecular forces, the thinner or less *viscous* the liquid. **Table 16.2** compares the different strengths of bonding interactions.

In the petroleum industry, the separation of different parts or fractions of naturally occurring crude oil allows the individual collection of many different products. The initial, thick, sticky crude tar is heated in a column to

Table 16.2 The strength of a liquid's bonds has a lot to do with the ability to flow.

	Strength of bonding interactions (kJ/mol)
Ionic bonding	100–1000
Covalent bonding	100–1000
Dipole–dipole intermolecular forces	0.1–10
Dispersion (London) forces	0.1–10
Hydrogen bonding	10–45

separate the different parts of the oil which boil at distinct temperatures. The heated products that rise up the column all have different lengths of carbon chains and can be pulled off as pure fractions. This is called *fractionation*. Methane (CH_4), propane (C_3H_8), and butane (C_4H_{10}) are purified separately of any impurities and then used as fuels. Gasoline and diesel fuels are separated as hydrocarbon products of 6–12 carbon atoms per molecule. Tar with hydrocarbons of 20–40 carbon atoms per molecule can be collected at a different time and level than propane, since the larger the molecule, the higher the boiling point.

Surface Tension

A strange thing happens at the surface of many liquids. The forces that hold molecules together pull down and to the sides, but there is no equal pull from above, where the surface is exposed to air. Then, the intermolecular forces from the rest of the liquid molecules flatten its shape when pulling from below.

The flattening allows the molecules to "float" or "ride" the molecules just below them that still keep their structure. Then rather than sinking, they form a film on the surface. When a molecule is in the middle of the liquid, it is attracted equally in all directions by the intermolecular forces. When a molecule is at the top, the forces underneath pull it unequally.

The stronger the molecular forces of a liquid, the greater the surface tension. Look at the strong intermolecular forces of water. When dropped as rain through the air, water droplets are spherical due to surface tension, and stretched slightly longer by gravity. Water droplets in microgravity experiments in space take on the shape of perfect spheres.

Vaporization

Liquids are affected by the amount of surface area exposed to air. Liquid surface molecules sometimes reach the energy needed to escape the rest of the liquid sample and become a vapor (gas). This is called *vaporization*.

> **Vaporization** is the way that molecules change from a solid or liquid to a vapor (gas).

In science fiction movies, the alien monster is sometimes "vaporized" by the hero's ray gun. This uses the same idea. The alien monster's molecules go from a solid threatening form to being scattered harmlessly into the air.

Vaporization needs heat to occur. Some liquids can go to the vapor form at room temperature and use heat from the environment. When water or perspiration dries (turns to vapor) from the surface of the skin, it uses body temperature. Heat energy from the body gives water molecules the energy to break surface tension attractions and become vapor. The amount of heat that it takes to vaporize 1 mol of liquid at a constant temperature and pressure is called the *molar heat of vaporization*.

Condensation

When a closed container is completely full of molecules in the vapor form above the surface, those molecules become jammed together. In a closed container, vaporization goes on only until the space above the liquid is *saturated*, or so full of the vaporized molecules of the liquid that there is no more room to expand.

When a container's air space becomes saturated, some of the vaporized molecules crash back into the liquid's surface and are captured. When this happens, the liquid form is preserved.

> **Condensation** is the opposite of vaporization. Molecules go from a vapor (gas) form back to a liquid state.

In a closed container, the rate of evaporation and condensation is not steady the whole time. It changes constantly. At first, the molecules slowly enter the vapor state. After a while, the closed air space is full of molecules and the liquid state grabs back the molecules that hit the surface. When this happens, the liquid sample has reached a state of *equilibrium*.

While the rates of vaporization and condensation are different at first, the rate of vaporization begins to slow down, while the rate of condensation begins to speed up. When the two rates become the same, *dynamic equilibrium* is reached. The exchange is dynamic because the molecules are not stuck, but continue to move back and forth between phases. The overall number of molecules in each phase is constant when the rates are the same.

> **Equilibrium** of a liquid in a closed environment takes place when the rate of condensation and evaporation is balanced. **Dynamic equilibrium** comes about when both forward and reverse reactions happen at the same time.

Equilibrium can be easily remembered in the following way:

$$\text{Liquid} \leftrightarrow \text{gas}$$

$$\text{Vaporization} \leftrightarrow \text{condensation}$$

When liquid molecules turn to vapor (gas) at equal rates or molecules vaporize at the same rate that they condense, then dynamic equilibrium is accomplished. This dynamic equilibrium can continue to change. When heat is applied, molecules get more energy and vaporize quicker, but at the same time, condensation is speeded up and equilibrium is established at a higher temperature. This is why it is called dynamic, because it continues to move.

Solubility

Solubility takes place when one compound is dissolved into another. These compounds, looking for others to bind to and become more stable, separate into individual ions. In general, the solubility of any solute is written as the ratio of grams of solute per 100 grams of water at a specified temperature.

Polar liquids, like water, are able to dissolve polar and ionic solutes. Non-polar liquids, like gasoline and acetone, are able to dissolve non-polar solutes.

Most chemical reactions are done in solutions using the different properties that a solution has apart from its component parts. Often, the melting point or freezing point of a solution is lower for the solution than its parent liquids. For example, ethylene glycol (CH_2OHCH_2OH) when mixed with water serves as an anti-freeze in motor vehicles. The combined solution of water and ethylene glycol freezes at a temperature below water's freezing point (–13°C instead of 0°C).

Evaporation

In an open container, vaporization is not limited by the amount of open space above the liquid. It can continue until all the molecules have gone into the air. When this happens, it is called *evaporation*. That is why fish tanks need to have water added every few days; the water molecules escape to the air and the liquid water level goes down.

Evaporation is different from vaporization in that it happens below the boiling point of a liquid, generally at room or outside temperatures.

Lakes and streams get smaller and smaller in hot weather if there is no rain. The water molecules evaporate into the atmosphere and don't come back down again. Water from underground reservoirs, called aquifers, are used up and not replenished. In a drought, water rationing is a way to limit water use until the environment changes, condensation happens, and water is returned to the earth.

Boiling Point

Temperature has a huge affect on solids and liquids. Since many elements are solid at room temperature (15–25°C), we normally think of them as solids. When solids are heated, they melt and become liquid. As we learned in Chapter 15, this temperature is called the melting point.

When we look at liquids, however, *boiling point* is like the melting point for solids. It is the specific point where enough heat energy has been added to the molecules of a liquid to allow it to vaporize and become a gas. This energy change is usually needed to cause the molecules to move from the in-between liquid form to the gas form.

> **Boiling point** is the temperature at which the vapor pressure equals atmospheric pressure. This allows liquid molecules to escape intermolecular bonding forces and escape into the air (as vapor).

Table 16.3 shows the big range of boiling points for liquids. It makes it easy to see how the millions of different reactions that happen with liquids are affected by temperature.

Table 16.3 Liquids have a wide range of boiling points depending on structure and bonding.

Name	Boiling point (1 atm pressure, °C)
Mercury	356.6
Water	100.0
Ammonia	−33.4
Ethyl alcohol	78.3
Vinegar	118.0
Acetone	56.2
Benzene	80.1
Iodine	184.0
Bleach (chlorine)	−34
Hydrochloric acid	109.0
Xylene	140.0

Liquids have a lot more freedom to perform different activities that denser solids cannot. In the next chapter, you will see how gases have even more exciting adventures.

Quiz 16

1. Density is
   (a) determined by the boiling point of the liquid
   (b) measured in grams per milliliter
   (c) the capability of a liquid to flow or not flow freely
   (d) the measurement of how much gas can be dissolved in the liquid

2. Relative density (specific gravity)
   (a) is the ratio of the density of a sample in liquid form divided by the density of the sample in solid form
   (b) is the ratio of the boiling point of a liquid at atmospheric pressure divided by the boiling point of a liquid at two atmospheres
   (c) measures volume of a sample when placed in water
   (d) is the ratio of the density of a sample at 20°C divided by the density of water at 4°C

3. Viscosity is the
   (a) temperature at which a liquid turns to a vapor (gas) at atmospheric pressure
   (b) temperature at which a vapor condenses into a liquid
   (c) capability of a liquid to flow or not flow freely at room temperature
   (d) ability of liquid molecules to turn to vapor

4. In the petroleum industry, the separation of different parts of naturally occurring crude oil and the collection of many products is an example of
   (a) vaporization
   (b) surface tension
   (c) condensation
   (d) fractionation

5. The stronger the molecular forces between molecules the
   (a) lower the boiling point of the liquid
   (b) more viscous a liquid
   (c) easier it is to condense
   (d) weaker the surface tension

6. Surface tension
   (a) is the force that pulls molecules down and to the sides
   (b) is the capability of a liquid to flow or not flow freely at room temperature
   (c) measures the evaporation rate of a liquid
   (d) is the surface of the liquid where evaporation takes place

7. Vaporization
   (a) is when a vapor turns to a liquid
   (b) is the capability of a liquid to flow or not flow freely at room temperature
   (c) is when a solid turns into a liquid
   (d) is the way that molecules change from a solid or liquid to a vapor

8. Boiling point
   (a) is the temperature at which the vapor pressure equals temperature
   (b) is the pressure at which mercury becomes a liquid
   (c) is the temperature at which the vapor pressure equals atmospheric pressure
   (d) is always longer when you are watching and waiting for it to boil

9. Solubility
   (a) takes place when one compound is dissolved into another
   (b) only happens with solutions containing acetone
   (c) of two or more mixing solutions cannot be performed
   (d) is a reaction that takes place in the sun

10. Dynamic equilibrium
    (a) occurs when both forward and reverse reactions happen at the same rate and time
    (b) is achieved only in open containers
    (c) occurs when a measuring cylinder is balanced between two glass rods
    (d) occurs when pressure and atmosphere are equal

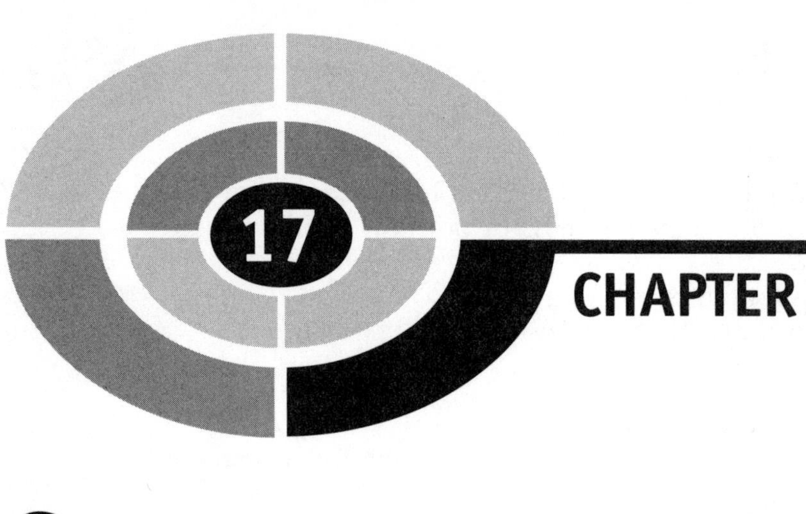

CHAPTER 17

Gases

Gases have always been a mystery. Ancient humans saw bubbles of gas form when they brewed ales and spiced ciders from grain and fermented fruit. Some tribes believed intestinal gases were somehow connected to the spirit. When the spirit was unhappy, ill humors (sickness) would plague a person with an excess of gas. When naturally occurring pockets of natural gas from the earth were discovered, it was thought the earth's spirit was releasing the errors of the people upon the land.

Until the 1700s, spontaneous combustion, the explosive occurrence of fire, was thought to be caused by mice, since the sudden combustion of stored grain in barns always seemed to happen when there were lots of mice around.

Modern chemists have learned a lot since then. With the development of precise equipment that can measure minute amounts of elements, accurate information about gases can now be gathered and studied.

The gas club is a lot more active than the liquid or solid clubs. Gases have livelier characters and no set boundaries. When allowed to escape from a container, they spread out into whatever space there is. If the space is the size of a room, they expand to fill the space. If the space is the outside atmosphere, they spread out infinitely, contained only by temperature changes and directed by wind currents.

Gases are the least compacted form of matter.

Gases are known for preferring to be as far as possible from each other, with no special shape or volume. Unlike solids and liquids, they are independent of one another.

Some common gases found in this free form club include nitrogen, oxygen, air, steam, carbon dioxide, helium, and argon. The air we breathe is a gas, except on very humid summer days when the air seems so loaded with water as to be nearly a liquid.

Table 17.1 gives some of the general characteristics of gases.

Table 17.1 Gases are the wild and carefree members of the three matter forms.

Gases
No set shape or volume
Expand to fill shape of container
Can be compressed by increasing pressure
Mix completely and spontaneously
Move constantly, quickly, and randomly
Smaller mass gases move more quickly than gases with larger masses
No strong molecular forces between particles
When particles collide, no energy is lost
All collisions are elastic
Low density

Atmosphere

Although it may seem like it, the air we breathe is not limitless. It reaches out about 30 kilometers from the surface of the Earth, but is only breathable to humans to about 14,000 feet or 4.3 kilometers. The air we breathe, in fact, is not just one gas, but several.

Outside and inside air is made up of roughly 78% nitrogen, 21% oxygen, and 1% argon with a smattering of 3–4% water vapor, carbon dioxide, sulfur dioxide, and the list goes on depending on where in the world you live.

Carbon Dioxide

Something important to consider is that the amount of polluting (chemicals not found in high levels in nature) gases in the atmosphere is rising. Levels at the times of the pyramids with fires and local industry, were around 80–100 parts per million (ppm), in 1900, the levels of CO_2 were less than 300 ppm. Today carbon dioxide levels approach 400 ppm. Since the human body does not do well breathing low levels of oxygen or pollutants, this is a significant problem. Not only is the air no longer pure in many parts of the world, but increasing levels of carbon dioxide add to the problem of rising global temperatures.

When gases expand and mix with other gases to fill available space, it is called *diffusion*. This is how environmentalists measure the levels of industrial gases in the air. When there is a gas release, they measure the amount of gas in parts per million or parts per billion. From these measurements, they can figure out the released amount and whether or not the diffused concentration is harmful to humans.

Kinetic Gas Theory

Gas molecules are always on the move. They are always bouncing off each other and other things like the walls of a container, or people, places, and things. They are super charged with energy. When scientists talk about this crazy motion of gases, they call it *kinetic energy*.

The kinetic energy of gases can be calculated. It is equal to one-half the mass (m) of the sample multiplied by the velocity squared (v^2):

$$\text{Kinetic energy} = {}^1\!/_2 m\, v^2$$

Then if a scientist has a sample of a known mass in a container with the molecules bouncing all over the place, the kinetic energy can be calculated using the equation above.

Chemists calculate the kinetic energy of gases based on their temperature, which in turn affects their velocities. The average kinetic energy of a gas molecule depends on the absolute temperature of the gas.

Gas Pressure

The atmosphere contains different gases as we learned earlier. These gas molecules collide with everything in our world, all the time. The Earth's gravity affects the force with which gas molecules hit people, objects, and each other. Gravity's pull on gas molecules decreases when molecules get farther and farther away from the Earth. Their weight is decreased without the constant tug of gravity although their mass remains the same. It is the same reason why astronauts are weightless in space and weigh only about a third of their weight on the moon (with a lot less gravity) as they would on Earth.

Atmospheric pressure is caused by the weight of the air per unit of area.

Since the first experiments to find this difference in pressure were performed with mercury and tall glass tubes by Italian scientist Evangelista Torricelli, the standard unit for pressure was called the *torr*. Experiments found that at sea level, the lowest land location to measure, that atmospheric pressure is equal to 1 atmosphere (1 atm).

$$1 \text{ atm} = 760 \text{ mm Hg} = 760 \text{ torr}$$

Note: When engineers and mechanics calculate pressure, they sometimes use the units pounds per square inch (psi).

Avogadro's Law

In 1811, Italian physicist Amedeo Avogadro noted that when you have equal volumes of gases at the same temperatures and pressures, the gases will have equal numbers of molecules. Avogadro found that this direct relationship between the number of molecules and the volume of the sample could be written in the following way: V/n (number of molecules) $= k$ (a constant)

$$V_1/n_1 = V_2/n_2$$

Temperature and pressure are unchanged and constant in Avogadro's law.

Avogadro's number (L) is a constant number of atoms, ions, or molecules in a sample. It is equal to the number of atoms in 12 grams of carbon-12 or 6.022×10^{23}.

To give you an idea of the unthinkable number of molecules contained in Avogadro's number, take a bunch of hazelnuts and cover the United States. To equal Avogadro's number, you would cover it with a layer over 100 kilometers (about 70 miles) deep.

Standard Temperature and Pressure

When studying the gas laws, you will sometimes see a problem talk about standard temperature and pressure (STP). **Table 17.2** lists the values for temperature and pressure that are called *gas standards*.

Table 17.2 Using standard temperature and pressure units for gases makes conversions simpler.

Standard temperature and pressure	(STP)
1 standard temperature	0°C
1 standard temperature	273 K
1 standard pressure	1 atm
1 standard pressure	760 torr
1 standard pressure	14.7 psi

Boyle's Law

In 1662, an Irish chemist named Robert Boyle tried to figure out how gas is affected by outside factors. To test this, he bent a glass tube into a hook shape and sealed one end. Into this tube he poured mercury. Boyle discovered that the mercury pushed ahead of it a small volume of air to the end that could not escape out of the tube. Since mercury is so heavy and dense, Boyle found that the more he poured into the tube, the harder it pushed against the trapped air at the end of the tube. After adding enough mercury to push (or compress) the trapped air into $^1/_2$ of its space (or volume) he realized that the more pressure pushing against a trapped volume of gas, the more it was compressed if the temperature stays the same. Boyle described pressure affect in the following way.

> In **Boyle's law**, when temperature is held constant, a volume of gas is inversely proportional to the pressure; $V \propto 1/P$

With further tests, Boyle found that by doubling the pressure, the volume of the gas was reduced by $^1/_2$. When the pressure was tripled, the volume of gas was squashed to $^1/_3$ of its original volume.

Boyle decided to multiply both sides of the equation by pressure (P) to get rid of the inverse $1/P$. Then, if you know two volumes and want to figure out the pressures or have two pressures and want to find the volumes, then the formula below will give the unknown values:

$$P_1V_1 = P_2V_2$$

EXAMPLE 17.1

See if you can figure out the amount of oxygen in a container if it has a volume of 4.0 liters. The pressure of the gas is 1470 psi when at 25 °C. If 1 atm of pressure is pushing on the volume, with no temperature change, find the volume of the oxygen.

$$P_1 = 1470 \text{ psi} \qquad P_2 = 14.7 \text{ psi}$$
$$V_1 = 4.0 \text{ liters} \qquad V_2 = X \text{ liters}$$
$$V_2 = 4.0 \text{ liters} \times 1470 \text{ psi}/14.7 \text{ psi} = 400.0 \text{ liters}$$

Since a greater volume of gas can be stored at a higher pressure, many gases are stored at increased pressure.

Charles' Law

The second of the ideal gas laws has to do with the affect of changing the temperatures of gases. This different angle of research was done in 1787 by French physicist Jacques Charles. Charles was said to be very interested in the hot air ballooning that was sweeping France as a huge sport at the time. He was known as one of the best balloonists in France. In fact, Charles was the first to use helium to inflate a balloon capable of carrying passengers.

Charles' scientific nature caused him to use ballooning as a way to test his ideas about gases and temperature. He found that the more a gas was heated, the more its volume increased. He described this with the following equation:

$$V_1/T_1 = V_2/T_2$$
$$V_2 = V_1 \times T_2/T_1$$

If we think of what is happening with the atoms of the gas, it is easy to remember Charles' law. Temperature (heat) provides energy to a sample. When atoms are energized (heated), they move around a lot. Like a happy puppy that can't stay in one place, heated atoms get crazy wild and hit the sides of their containers harder and more often causing it to expand.

> In **Charles' law**, when pressure is held constant, a volume of gas is directly proportional to the Kelvin temperature; $V \propto T$

The kinetic energy (KE) increases, but the mass stays the same, so the velocity has to increase.

$$\textbf{\textit{Kinetic energy}} = {}^1\!/_2 m\, v^2$$

EXAMPLE 17.2

If we have a balloon of 1.0 liter that is flexible and can expand with increased temperature (from 20°C to 45°C), find the volume of the balloon after it is heated. (Hint: add 273 to the Celsius temperatures to get everything into Kelvin.)

$$T_1 = 20 + 273 = 293 \text{ K}$$
$$T_2 = 45 + 273 = 318 \text{ K}$$
$$V_1 = 1.0 \text{ liter},\ V_2 = X$$
$$1.0 \text{ liters}/293 \text{ K} = x \text{ liters}/318 \text{ K}$$
$$x = 1.0 \text{ liter} \times 318 \text{ K}/293 \text{ K}$$
$$x = 1.09 \text{ liters} = V_2$$

Gay-Lussac's Law

Around the same time that Charles was ballooning and experimenting in France, another French scientist, Joseph Gay-Lussac, was studying the connection between temperature and gas pressure. His research added the third of the ideal gas laws. Gay-Lussac discovered that as temperature increases and kinetic energy increases, pressure increases too.

> In **Gay-Lussac's law**, when volume is held constant, the pressure of a gas is directly proportional to the Kelvin temperature; $P \propto T$.

It's a case of atoms dancing wildly in a constant volume again. As the temperature increases, the pressure increases and the atoms collide with the container's walls faster and harder. This increases the kinetic energy. The following equation describes what happens:

$$P_1/T_1 = P_2/T_2$$
$$P_2 = P_1 \times T_2/T_1 \text{ or } T_2 = T_1 \times P_2/P_1$$

When you read the label of a pressurized spray paint can, you will probably see a warning not to let the can come in contact with heat. You can thank Gay-Lussac for this warning. If the can is heated enough, the pressure will increase and the can will explode. Besides being very dangerous, you will paint everything in sight. The take home chemistry message is, *never heat a spray can*!

EXAMPLE 17.3

If you have a pressurized (875 torr), room temperature (25°C) hair spray can with a volume of 15 ounces (oz.) that is in a house fire and heated to 1500°C, find the pressure inside the can before it explodes.

$$V_1 = 15 \text{ oz.}, V_2 = 15 \text{ oz. (just before it explodes)}$$
$$T_1 = 27°\text{C}, T_2 = 1500°\text{C}$$

Remember to add 273 to get the temperature in kelvin.

$$T_1 = 27 + 273 = 300 \text{ K}$$
$$T_2 = 1500 + 273 = 1773 \text{ K}$$
$$P_1 = 875 \text{ torr}$$
$$P_2 = x \text{ torr}$$
$$P_2 = 875 \text{ torr} \times 1773 \text{ K}/300 \text{ K} = x \text{ torr}$$
$$P_2 = 5171 \text{ torr}$$

The pressure at this high heat is nearly 6 times what the spray can is designed to hold!

Ideal Gas Law

These three gas laws are referred to as the *ideal gas laws*. They were discovered by different scientists at different times, but all add up to explain the strange and amazing things that gases do in different conditions. The formula that considers all three laws is written as:

$$PV = n\ RT$$

where P = pressure, V = volume, n = number of moles of gas at constant pressure and temperature, R = molar gas constant (0.821 liter × atm/ (kelvin × mol)), T = temperature. An *ideal gas* is one that meets all the rules of the gas laws. Gases that are mixtures of different molecules have some quirks that don't follow the gas laws exactly.

Dalton's Law of Partial Pressures

John Dalton that we learned about in Chapter 1 as the father of the atomic theory, came up with an idea about how gas pressure works. Like Gay-Lussac, Dalton had a hobby. He was interested in meteorology; the study of the weather. While studying changes in the weather, Dalton did some experiments with vapor pressure. What he found was that, like people, gases are unique and behave in their own way when in a mixture. Each gas, for example, compresses at its own pressure. When using three different gases at a constant temperature, Dalton found that the total pressure of the three gases was equal to the sum of each of the three individual gases. This general rule became known as Dalton's law of partial pressures.

Dalton's law of partial pressures says that when you have more than one gas mixed with one or more different gases, the pressures of each gas will add together to give the total pressure of the mixture.

Dalton's law is probably the easiest of all the gas laws to remember. It is written like this:

$$P_{\text{total}} = P_1 + P_2 + P_3 + P_4 + P_5 + \cdots$$

EXAMPLE 17.5
If you have an atmosphere of mixed gases on an alien planet, composed of oxygen ($p = 0.3$ atm), argon ($p = 0.1$ atm), nitrogen ($p = 0.8$), and neon ($p = 0.01$), what is the total pressure of the atmospheric gases?

$$P_{\text{total}} = P_1(0.3) + P_2(0.1) + P_3(0.8) + P_4(0.01)$$
$$P_{\text{total}} = 1.21 \text{ atm}$$

The total pressure of the alien atmosphere is greater than that of any of the single gases alone.

Combined Gas Law

Considering all the gas laws, then, you can figure out nearly every temperature, volume, and pressure of a gas if you know the other constants. This can be thought of as the *combined gas law*.

> The **combined gas law** is made up of the general rules for temperature, volume, and pressure described in Boyle's, Charles', and Gay-Lussac's laws.

It is written as:

$$P_1V_1/T_1 = P_2V_2/T_2$$

When you know five of the six values, you can figure out the missing value. If the volumes don't change, then you can write the combined gas law as:

$$V_2 = V_1 \times P_1/P_2 \times T_2/T_1$$
$$V_2 = V_1$$
$$1 = P_1/P_2 \times T_2/T_1, \text{ or } P_2/P_1 = T_2/T_1 \text{ (which is Gay-Lussac's law)}$$

EXAMPLE 17.6
If a bicycle tire has a volume of 0.5 m^3 at 20°C and 1 atm (760 torr), figure out the volume when taken into the mountains where the pressure is 720 torr and temperature is 14°C.

$$V_1 = 0.5\ \text{m}^3,\ T_1 = 20°\text{C},\ T_2 = 14°\text{C},\ P_1 = 760\ \text{torr},\ P_2 = 720\ \text{torr}$$
$$V_2 = x$$
$$T_1 = 20 + 273 = 293\ \text{K},\ T_2 = 14 + 273 = 287\ \text{K}$$
$$V_2 = 0.5\ \text{m}^3 \times 760\ \text{torr}/720\ \text{torr} \times 287°\text{C}/293°\text{C}$$
$$0.5\ \text{m}^3 \times 1.056\ \text{torr} \times 0.980°\text{C}$$
$$V_2 = 0.52\ \text{m}^3$$

As the temperature and pressure decrease, the tire volume is less compressed and feels mushy. To get the same riding feel on the tires, you might want to pump up the tires a bit.

Vapor Pressure

Since some gases are non-polar and don't mix with water, they can be collected using water as a filter. Water is displaced by the gas that is bubbled through it and collected. The vapor pressure of a gas is directly affected by temperature. As temperature increases, vapor pressure increases.

> **Vapor pressure** of a sample is equal to the partial pressure of the gas molecules above the liquid phase of the sample added to the pressure of the water vapor.

$$P_{\text{total}} = P_{\text{gas collected}} + P_{\text{water pressure}}$$
$$P_{\text{gas collected}} = P_{\text{total}} - P_{\text{water pressure}}$$

Molar Volume

To figure out the volume of a mole of gas, use Avogadro's number of 6.022×10^{23} molecules in a mole. To figure out the molar volume of a gas, multiply the density times its molar mass.

> **Molar volume** describes the density of a gas times its molar mass. At STP all gases have the same molar volume.

$$\text{Molar volume} = d \times m$$

The density of oxygen at STP is 1.43 grams/liter and its molar mass is 32.0 grams/mol. To figure out the molar volume of oxygen at STP, use the following formula:

$$32.0 \text{ grams/mol} \times 1 \text{ liter}/1.43 \text{ grams} = 22.4 \text{ liters/mol}$$

Since Avogadro's number is so huge, every gas works out to be very close to 22.4 liters/mol, so it is commonly said that the molar volume of every gas at STP is 22.4 liters/mol.

The activity of gases can be calculated using a lot of different ratios. If you know the temperature, pressure, and volume of gases, then lots of different ideas can be tested. The gas club is definitely the most changeable of the solid, liquid, and gas forms of matter.

Quiz 17

1. Which of the following is not a true statement about gases?
   (a) they are the least compacted form of matter
   (b) they are more active than liquid
   (c) they prefer to be as far as possible from each other
   (d) some gases take on very distinct shapes

2. Which of the following is not true of kinetic energy?
   (a) it is the type of energy a gas uses to stay in motion
   (b) it is named after the scientist Sorensen Kinet
   (c) it can be very easily calculated
   (d) kinetic energy = $\frac{1}{2}mv^2$

3. Boyle's law describes
   (a) the ideal gas law
   (b) the relationship of atmospheric pressure and temperature
   (c) describes when temperature is held constant, a volume of gas is inversely proportional to the pressure
   (d) an idea that does not apply to ideal gas laws

4. Charles' law explains
   (a) how gases are always on the move
   (b) how equal volumes of gases at the same temperature and pressure have equal numbers of molecules
   (c) the relationship of gas volume and gas pressure
   (d) when pressure is held constant, a volume of gas is directly proportional to the Kelvin temperature

5. Gay-Lussac's law explains
   (a) when volume is held constant, the pressure of a gas is directly proportional to the Kelvin temperature
   (b) when pressure is held constant, the volume of a gas is directly proportional to the Kelvin temperature
   (c) when temperature is held constant, the volume of a gas is inversely proportional to the pressure
   (d) the effect of changing temperatures on gases

6. Dalton's law of partial pressures states that when
   (a) temperature is constant and the gas volume expands, pressure equals a portion of the original
   (b) more than one gas mixes with one or more different gases, the pressure of each gas will add together to give the total pressure of the mixture
   (c) more than one gas mixes with one or more different gases, the total pressure of the mixture will be the same as the heaviest gas
   (d) gases come together, only two will combine at any one time

7. Which of the following is not true about atmospheric pressure?
   (a) it stays the same regardless of elevation
   (b) it is caused by the force of the air molecules that push against a unit of area
   (c) the standard unit of pressure is called a torr
   (d) 1 atmosphere is found at sea level

8. When gases expand and mix with other gases to fill available space, it is called
   (a) vaporization
   (b) evaporation
   (c) solidification
   (d) diffusion

9. In the equation $PV = n\ RT$, n is equal to
   (a) pressure
   (b) volume
   (c) number of moles of gas
   (d) temperature

10. The pressure of a gas is directly proportional to the Kelvin temperature in
    (a) Boyle's law
    (b) Gay-Lussac's law
    (c) Boyd's law
    (d) Charles' law

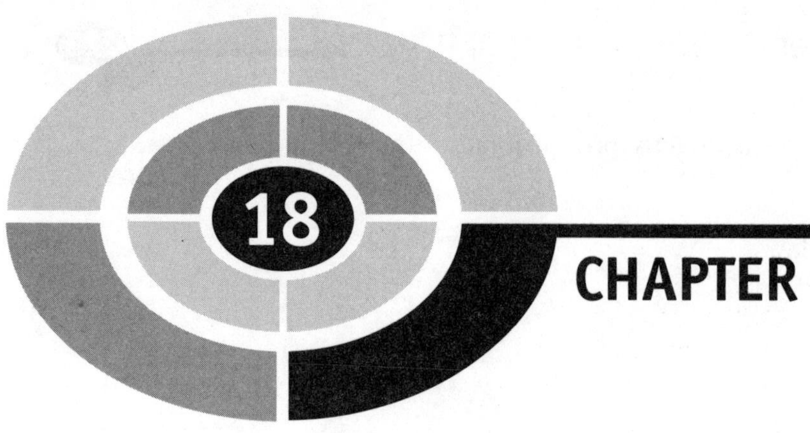

Biochemistry, Nanotechnology, and the Future

Chemistry is life! This should be everywhere on shirts and billboards! Chemistry describes the basic building blocks of matter. Atoms, molecules, sub-atomic particles, solids, liquids, and gases make up everything we know to exist in our world.

The fact that chemistry plays a big role in our lives is good news for people interested in working in chemistry or in fields that make use of specific element interactions. There are nearly as many practical applications as fields of study. Let's look at a few of them.

Biochemistry

The word *biochemistry* describes the chemistry of living systems. Some of these were described when we talked about organic chemistry and the molecules that make up living things.

One of the ways that changes take place in organic molecules is through the life cycle of microorganisms. A single-celled organism, through its *metabolism,* builds up or breaks down organic molecules.

The biological molecules that make up living cells, organs, systems, and the environment can be divided into four types: *proteins*, *carbohydrates*, *nucleic acids*, and *lipids*. Most of these molecules are simple structures covalently bonded to similar molecules, but some can reach incredible sizes in molecular terms. They are called *macromolecules*. **Figure 18.1** shows an example of a macromolecule: β-carotene (the yellow color in carrots).

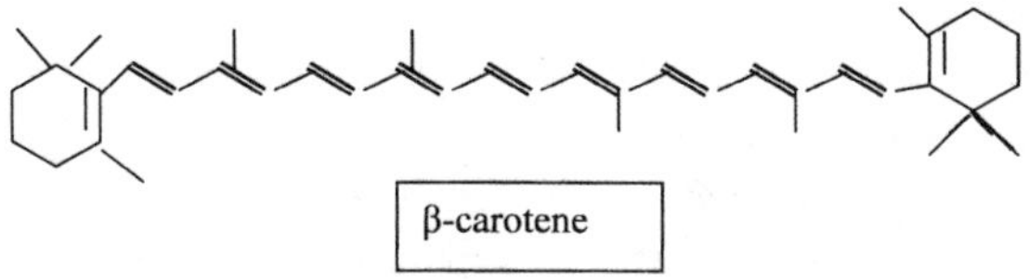

β-carotene

Fig. 18.1. The structure of β-carotene gives an idea of the complex nature of proteins.

Proteins

Organic protein molecules serve different functions for living systems. Some offer structural strength, as in the chitin shells of crabs and bone of mammals, some provide transport, as in hemoglobin, some act as blueprints for cell and organ development (DNA), some serve as messengers (hormones) between body organs, and some speed up metabolic reactions (enzymes).

The molecular weight of proteins ranges from 6×10^3 to millions of atomic mass units.

> **Proteins** are made up of small molecules that contain an amino group ($-NH_2$) and a carboxyl ($-COOH$) group. These molecules are called **amino acids**.

Most protein reactions are made up of many different combinations of amino acids reacting with water, salts, and other elements to create or enhance needed functions. Amino acids can contain a variety of non-protein ions like some of the metals (Zn^{2+}, Fe^{2+}, Mg^{2+}). For example, the hemoglobin molecule uses iron as a critical part of its function of transferring oxygen within living systems. Amino acids are bonded by *peptide* (C–N) *bonds*.

In 1908, a German chemist and medical researcher, Paul Ehrlich, working with aniline dyes in the staining of disease-causing microorganisms discovered that these chemical solutions could also kill the organisms without killing the patient. He shared the Nobel Prize for Medicine with Elie Metchnikoff in 1908 for his work. Two years later, Ehrlich developed the first antibacterial agent, salvarsan, to treat syphilis. Because of his interest in treating diseases with chemical cures, he became known as the *father of chemotherapy*.

There are many other areas where biochemistry led to important applications. In 1993, Kary Mullis received the Nobel Prize for Chemistry for the invention of a polymerase chain reaction technique for amplifying deoxyribonucleic acid (DNA). That same year, Canadian chemist Michael Smith also received the Nobel Prize for Chemistry for his technique of splicing foreign gene segments, designed to modify the production of a specific protein, into another organism's DNA. This opened the gates to a flood of research on designer proteins and molecules produced for a specific purpose.

Designed proteins are now being used for everything from better medicines, like insulin to treat diabetes and artificial fabrics to treat and protect burn patients, to industrial foams that clump and eliminate spills from oil tankers and medicines to counteract biological poisons.

Environment

For centuries, the environment was so vast and scarcely populated that to humans it was limitless. Wood was used freely, refuse was left wherever convenient, and as long as you are upstream, you could dump whatever you wanted into the rivers and the ocean. Now due largely to better medicines and health care, people are living to their seventh, eighth, and ninth decades. Entire populations are no longer getting wiped out by disease.

Scientists began to look at humankind's impact on this planet as the world's population swelled to 6 billion and more. Chemists are becoming mystery investigators. The environment is a very complex mixture of elements with different concentration spikes in many areas. Industrial cities have higher levels of metals and acids in their air than rural countryside areas. Scientists must work together to combine all available information from air and water samples as well as those from industrial emissions in order to piece together the puzzle of total environmental impact. The interconnectedness of all forms of life also affects the complexity of environmental pollution.

In 1995, three chemists, Mario Molina, Sherman Rowland, and Paul Crutzen, warned world leaders of damage being done to the O_3 (ozone) layer. This natural layer of O_3 molecules, located from 9 to 30 miles up into the atmosphere, protects the Earth from cancer-causing and damaging ultraviolet radiation from the sun. They discovered that human-made compounds of nitrogen oxides and chlorofluorocarbon (CFC) gases, used as refrigerants and propellants in spray cans, reacted with atmospheric ozone and reduced it. For their work, they received the 1995 Nobel Prize for Chemistry.

In response to the ozone depletion problem, chemists began looking for replacement refrigerants that didn't affect ozone. Substitutes were found and the environmental problem lessened.

Many of the elements used today were discovered using cutting-edge technology and equipment. Since the 1960s, many of the elements added to the Periodic Table were human-made and not found in nature. These molecules have unheard of uses that many research and applications chemists and biochemists are just beginning to understand.

Chemists working in the plastics industry came under heavy criticism when landfills became overloaded with the new, disposable containers of plastic and a softer compound called Styrofoam. Environmentalists sounded the alarm for consumers to think before they bought products, especially fast food, that came in these containers.

In order to meet the new concern, chemists doubled their interest in the biodegradability of plastic products. They found that the addition of complex *carbohydrates* (polysaccharides) to plastics allowed microorganisms to break down the plastic products.

> Molecules that can be broken down into simpler elements by microorganisms are called **biodegradable**.

> **Carbohydrates** make up a large group of organic compounds containing carbon, oxygen, and hydrogen.

Carbohydrates have the general formula of $C_x(H_2O)_y$. There are three main groups of carbohydrates. The first are the *simple sugars* or *monosaccharides*. Some of these are the simple fruit sugars, fructose and glucose, with the formula $C_6H_{12}O_6$. The simple milk sugar that many people with milk sensitivities have trouble with is lactose. The second group is known as the *complex sugars* or *disaccharides*. These are combined sugars that make up honey and table sugar, sucrose and maltose ($C_{12}H_{22}O_{11}$). Complex carbohydrates with complicated, folded structures make up the starch added to plastics, as well as cellulose of plant cell walls and rayon (processed cellulose). They have the formula $(C_6H_{10}O_5)_n$ where n is an extremely large number. These are commonly called macromolecules because of the number of elements and huge size compared to simple molecules. **Figure 18.2** shows the structure of glucose and cellulose.

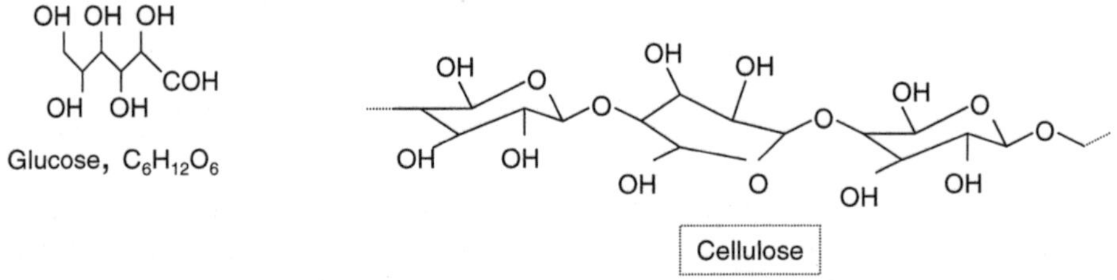

OH OH OH
COH
OH OH
Glucose, $C_6H_{12}O_6$

OH O OH O OH OH O OH O
OH OH OH O OH OH
Cellulose

Fig. 18.2. Carbohydrates like glucose and cellulose serve many functions in organic organisms

Radioactive Waste

As we saw in Chapter 11, one major drawback to most of the radioactive elements discovered and produced in greater than the extremely small amounts found in nature is that they accumulate in the environment. Land, water, and air are affected by radioactive contamination. Depending on the wind or water flow, radioactive levels remain in place or are spread over a wide region. Different elements have very different *decay rates*.

> **Radioactive decay** occurs when certain element isotopes are lost and there is a release of energy in the form of radiation (alpha and beta particles and gamma rays).

The three main types of radiation given off during the breakdown of radioactive elements are *alpha* (α) and *beta* (β) particles, and *gamma* (γ) rays. Gamma rays are high-energy electromagnetic waves like light, but with a shorter, more penetrating wavelength. Though alpha and beta particles are dangerous to living things since they penetrate cells and damage proteins, gamma rays are much more penetrating and harmful, stopped only by thick, dense metals like lead.

The waste produced in different forms of matter transformation must eventually be broken down. This is an area of ongoing concern and study for many governments who are trying to figure out how to dispose of radioactive wastes from nuclear power plants and atomic weapons. Protecting their populations from handling accidents or terrorist nuclear threats will continue to promote research in understanding the reactivity and degradation of radioactive compounds and elements.

Physical Chemistry

Research started from the middle of the 1900s and on pointed toward particles of even smaller dimensions than atoms, called *hadrons* and *leptons*.

Hadrons are made up of *baryons* and *mesons*. Baryons are made up of protons, neutrons, and other short-lived particles. Mesons are made up of *pions*, *kaons*, and other short-lived particles. Leptons are made up of electrons and different types of neutrinos (*tau* and *muon*). You can't really see or measure any of these, but only record their effects. It is a lot like the wind. Leaves can blow around like crazy, but unless there is dust in the air, you can't see exactly what is happening.

Hadrons are also made up of even tinier particles called *quarks*. Six kinds of quarks have been described. They are: *up*, *down*, *charm*, *strange*, *top*, and *bottom*. Each has a different charge and there are also *anti-quarks* with the opposite charges of their twin.

Strange and charming molecular chemists work with protons, electrons, and neutrons as we have seen throughout our study of chemistry. Quarks and quark theory occupy the thoughts of the theoretical physicist. These scientists search for insight into nuclear binding energy, the energy that keeps various nucleons together in the nucleus, and how everything comes together in the greater picture of chemical interactions.

Nanotechnology

The subject of the tsunami (really big wave) of current scientific research is in the area of *nanotechnology*. An entire special issue, the September 2001 issue of *Scientific American*, was devoted entirely to the topic. The "big guns" in the field described their work in the areas of Medical Nanoprobes, Buckytube Electronics, Living Machinery, Atom-Moving Tools, New Laws of Physics, and Nano Science Fiction. The cover story of *Scientific American*'s January 2003 issue describes "The Nanodrive." Through the use of individual silicon molecules and etching onto a special polymer medium, computer drives of the future will process and perform data storage tasks of several gigabytes of information on a chip the size of a postage stamp.

> **Nanotechnology** is the study of elements at the single atom level or 10^{-9} (1 billionth of a meter) scale.

To put the nanometer scale into everyday measurements, think of the size of a person 2 meters tall (about 6 feet). A gnat, 2 millimeters long is 1000 times smaller than a person 2 meters tall. One cell in a gnat's body has a nucleus of about 2 micrometers long or 1000 times smaller than the size of the entire gnat's body. A nanoscale molecule is roughly 2–10 nanometers long or 1000 times smaller than the length of the nucleus of a gnat's cell.

Nanomolecules are *super* small!

American molecular chemist, Richard Smalley, at Rice University in Houston, Texas, studies atoms and molecules at the nanomolecular level. His research with soccer ball-shaped carbon molecules led him to receive the 1996 Nobel Prize for Chemistry along with Robert Curl, Jr. and Harold W. Kroto for the discovery of fullerenes (C_{60}). Smalley's current research is directed toward fullerene nanostructures and involves the investigation of carbon single-wall *nanotubes*, nanoscale tubular structures built of graphene sheets (fullerenes). His research is geared toward the development, production, characterization, and use of tubular fullerene molecules, nanotube single-crystal growth, nanotube fibers, and other nanotechnology materials and applications.

Molecular Electronics

Currently the "star" of worldwide nanotechnology attention is focused on molecular electronics. Dr. James M. Tour, head of the molecular electronics effort at the Center for Nanoscale Science and Technology at Rice University, whose work focuses on the super small world of nanotechnology, has proposed experiments in which computer electronics are built from the "bottom up," molecule by molecule. Bottom up nanoscale construction is patterned after nature, with molecules forming cells that form tissues that form organs that form systems that finally form a total organism or person.

> **Molecular electronics** uses individual molecules or very small groups of molecules (carbon, oxygen, hydrogen, and nitrogen) to serve as transistors, conductors, and other electrical parts of computers and circuits.

Current computer processing "chips" are built from the "top down" in incredibly clean (dust-free) environments onto etched gold plates.

Tour's scientific research areas include molecular electronics, chemical self-assembly, carbon nanotube modification and composite formation, and synthesis of molecular motors and *nanotrucks* (molecules that bind and transport other molecules back and forth a short distance using an external electrical field) to name a few.

Tour has also attracted the interest of the National Science Foundation and others in his mission to teach chemistry, physics, biology, and materials science at the nanomolecular level through the animated adventures of actual molecules chemically synthesized in his Rice University laboratory. This project, called NanoKids™, seeks to open up science with computer animation, music, and a MTV style and make nanoscale science learning fun and simple.

Nanotechnology research has gained the attention of international companies like IBM and Microsoft. Computer circuitry built 100 times smaller than the tiniest components today seems almost miraculous. Molecular chemists are talking about building with atoms and molecules that are not visible with the strongest microscopes.

So how do we know this can be done? Simple chemical interactions and known properties of the elements are used. In fact, most nanomolecules are carbon-based polymers that are very similar to naturally occurring substances that handle electrical impulses all the time, like brain neurons.

Similar to their transferal of electrical impulses in the body, individual molecules alternate between two forms. They act like molecular on/off light switches and can store information or pass it along in a split second or less.

Research is being done to build nanodrives that write, read, and erase data using atoms and molecules, would hold several gigabytes of data and fit on the top of one key from a computer keyboard.

The future of chemistry is as big as our planet and as small as nanomolecules that can never be seen by the human eye. Research and applications of chemistry that consider everything from the purification of our air, to the speed of our computers will be increasingly important for decades to come. The materials and machines of science fiction will become reality as humankind grows in its knowledge of this fascinating science.

Quiz 18

1. Single-celled organisms break down organic molecules through
   (a) vaporization
   (b) cell division
   (c) metabolism
   (d) evaporation

2. Organic protein molecules serve living systems in which of the following ways?
   (a) strength
   (b) transport
   (c) messengers
   (d) all of the above

3. Paul Ehrlich
   (a) received the Nobel prize in 1948 for his treatment of mad cow disease
   (b) is known as the father of chemotherapy
   (c) discovered chemical solutions that kill microorganisms and patients
   (d) worked exclusively with chitin protein in crab shells

4. The core of the hemoglobin molecule is
   (a) iron
   (b) gold

(c) zinc
(d) magnesium

5. Macromolecules are
(a) ionic compounds found in minerals
(b) covalently bonded molecules of large size
(c) about 2 millimeters long
(d) can be found in mercury solutions

6. Amino acids are
(a) based on ethyl subgroups
(b) bonded by (OH–) bonds
(c) available as supplements to improve eyesight
(d) bonded by peptide (C–N) bonds

7. Ozone is
(a) made up of NH_3 molecules
(b) increasing all the time
(c) made up of O_3 molecules
(d) located 250 miles up in the atmosphere

8. The three main types of radiation given off during decay are
(a) neutrino waves
(b) alpha, beta, and gamma rays
(c) tachyon emissions
(d) alpha, beta, and zeta rays

9. Nanotechnology is the study of elements
(a) at the single atom level of 10^{-6} meters
(b) found outside a cell nucleus
(c) found only in platinum samples of high density
(d) at the single atom level of 10^{-9} meters

10. When molecules can be broken down, they are
(a) biodegradable
(b) catalytic
(c) vaporized
(d) saturated

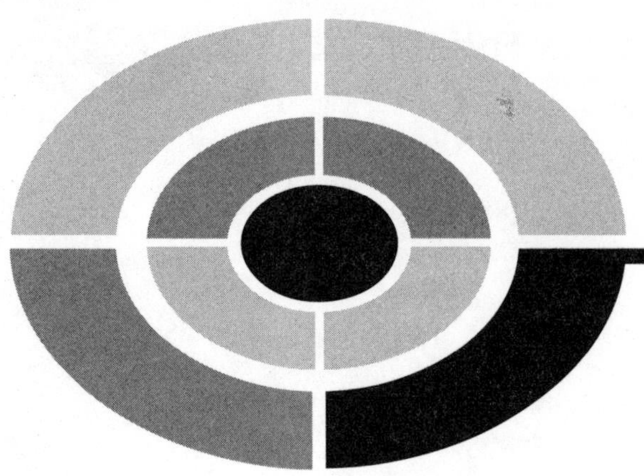

Test: Part Four

1. Which of the following is not a general rule in the laboratory?
   (a) Do not run your fingers through the burner flame
   (b) Do not eat or drink anything in the lab
   (c) Wash with a strong base before every experiment
   (d) Always wear goggles when using chemicals

2. Acetum is the Greek word for
   (a) hot dog
   (b) smelly feet
   (c) gladiator
   (d) vinegar

3. When ions are held together by electrical attraction, it is sometimes called
   (a) sticky element binding
   (b) electrostatic attraction
   (c) unilateral attraction
   (d) phosphorescence

4. In sodium chloride (NaCl) an ionic bond is formed when
   (a) the sun is directly overhead
   (b) the transfer of sodium electrons to the chlorine ion occurs

(c) crystalline structure is planar in form.
(d) the transfer of chlorine protons to sodium occurs

5. What is the easiest way to identify a sloppy chemistry student in the lab?
(a) look for someone who always wears red
(b) look for someone carrying around models of the human eye
(c) look for someone with a really studious expression
(d) look for someone with holes in his clothes

6. The electronegativity value of indium is
(a) 0.7
(b) 1.2
(c) 1.7
(d) 2.5

7. Chlorobromoiodomethane is an example of a
(a) chiral molecule
(b) chlorofluorcarbon
(c) soap
(d) transition metal

8. Isomers that are mirror images of each other and superimposable are called
(a) chiral
(b) enantiomers
(c) electron pairs
(d) elastomers

9. How many valence electrons does carbon have?
(a) 2
(b) 4
(c) 6
(d) 8

10. The unequal sharing of electrons between two atoms is called
(a) a dipole
(b) unfair
(c) may pole
(d) achiral

11. Chemical bonds are made through the interaction of
(a) neutrons
(b) orlons

(c) electrons
(d) mylons

12. *Metallic* bonds form between
(a) non-metal atoms
(b) noble gases
(c) a non-metal and a halogen
(d) metal atoms

13. Organic chemistry is the chemistry of
(a) nitrogen
(b) carbon
(c) sodium
(d) oxygen

14. The density of water is
(a) 1.00 g/mL at 4°C
(b) 1.00 g/mL at 10°C
(c) 1.00 g/mL at 40°C
(d) 1.00 g/mL at 45°C

15. The word *chiral* comes from the Greek word for
(a) chair
(b) hand
(c) water
(d) calcium

16. Isomers have the same molecular formulas, but different
(a) one-dimensional structures
(b) two-dimensional structures
(c) three-dimensional structures
(d) four-dimensional structures

17. Lava, a cold mountain stream, and mercury are all
(a) very thick
(b) red
(c) liquids
(d) metals

18. Relative density is also known as
(a) specific gravity
(b) non-specific gravity
(c) microgravity
(d) relative resolution

19. The size, strength and shape of molecules, along with intermolecular forces
    (a) are really nothing to worry about in chemistry
    (b) are important only in calculating atomic number
    (c) have no affect at all on bonding properties
    (d) have a big affect on the viscosity of liquids

20. There are two basic types of isomers, structural and
    (a) audio isomers
    (b) stereoisomers
    (c) magnetic isomers
    (d) ambivalent isomers

21. The stronger the molecular forces of a liquid, the
    (a) greater the surface tension
    (b) weaker the surface tension
    (c) lack of any surface tension
    (d) greater the chance for radioactivity

22. Up to 1700's, spontaneous combustion of grain was thought to be caused
    (a) by crazed pyrotechnicians
    (b) by mold
    (c) by mice
    (d) by bad barley

23. Gases are the
    (a) least compacted form of matter
    (b) easiest form of matter to measure when pouring
    (c) most compacted form of matter
    (d) are usually brittle

24. The constant number of atoms, ions, or molecules in a sample is known as
    (a) Doyle's Number
    (b) Avogadro's Number
    (c) Gasper's Number
    (d) Mozart's Number

25. Carbon dioxide levels today are around
    (a) 100 ppm
    (b) 200 ppm
    (c) 300 ppm
    (d) 400 ppm

26. A very important lesson we get from Gay-Lussac's Law is
    (a) to always use a graphing calculator
    (b) wear open shoes when working in the lab
    (c) to never heat a spray can
    (d) that pressure does not relate to temperature

27. The hemoglobin molecule uses what metal to transport oxygen?
    a) lead
    b) iron
    c) tin
    d) silver

28. Protein molecules that reach very large sizes are called
    a) nanomolecules
    b) linebackers
    c) blimp molecules
    d) macromolecules

29. The standard formula to calculate kinetic energy is
    (a) $KE = mv^2$
    (b) $KE = \frac{1}{2}v^2$
    (c) $KE = \frac{1}{2}mv^2$
    (d) $KE = mv$

30. The eletronegativity value of fluorine is
    (a) 1.4
    (b) 1.9
    (c) 3.2
    (d) 4.0

31. Atmospheric pressure is caused by
    (a) the weight of the air per unit of area
    (b) your peers
    (c) cosmic particles from outer space
    (d) the humidity, not the heat

32. Amino acids contain an amino group ($-NH_2$) and a
    (a) iodo group
    (b) carboxyl group
    (c) butyl group
    (d) cupric group

33. Gas molecules are
    (a) seldom moving at all
    (b) always on the move
    (c) packed densely together in the air
    (d) something to be avoided at a party

34. Who discovered the antibacterial agent, salvarsan?
    (a) Elisabeth Fleming
    (b) Linus Pauling
    (c) Paul Erhlich
    (d) Robert McKenna

35. Plant cell walls are made up of
    (a) hemoglobin molecules
    (b) noble gases
    (c) alkaline metals
    (d) complex carbohydrates with complicated, folded structures

36. Covalent bonds between atoms of the same elements are known as
    (a) convenient bonds
    (b) polar bonds
    (c) non-polar covalent bonds
    (d) citrus bonds

37. Nitrogen oxides and what other group of gases were found to react with atmospheric ozone and reduce it?
    (a) amino acid
    (b) zinc oxide
    (c) chlorofluorocarbon
    (d) green argon

38. Carbohydrates have the general formula of
    (a) $C_x(H_2O)_y$
    (b) $C(H_2O)$
    (c) $C(H_3O)_y$
    (d) $C_x(O_2)_y$

39. Which of the following is known as the Ideal Gas Law?
    (a) $E = mc^2$
    (b) $PV = n\ RT$
    (c) $P_{total} = P_1+P_2+P_3$
    (d) $P_1 = ET$

40. Radioactive decay occurs as a release of energy in the form of
    (a) alpha, beta, and wonka particles
    (b) alpha, beta, and gamma particles
    (c) beta and zuma particles
    (d) grandma particles

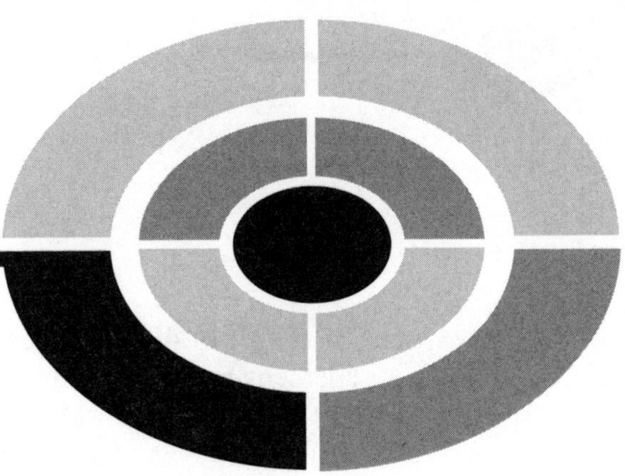

Final Exam

1. Which of the following elements is not naturally found in the human body?
(a) manganese
(b) sulfur
(c) iron
(d) thorium

2. This important tool provides the most information in one place on the elements
(a) the Periodic Table
(b) the octave rule
(c) chemical nomenclature
(d) the Bernouli rule

3. Rust is a material that is formed when
(a) copper is exposed to air and moisture
(b) iron is exposed to oil
(c) titanium is exposed to oil
(d) iron is exposed to air and moisture

4. In Australia the temperature in January is often 24°C, what is it in Fahrenheit?

(a) 90
(b) 94
(c) 97
(d) 101

5. If 12 out of 48 songs become number 1 hits in a 6 month period, what percent reach number 1 in that period?
(a) 15%
(b) 20%
(c) 25%
(d) 32%

6. Hadrons are made up of
(a) mesons and orlons
(b) baryons and nylons
(c) baryons and mesons
(d) mesons and quarons

7. Arsenic sulfide was added to whiten copper by alchemists, this step was called
(a) melanosis
(b) leucosis
(c) cystosis
(d) xanthosis

8. Nano is the Greek prefix for
(a) 10^2
(b) 10^6
(c) 10^9
(d) 10^{12}

9. Color, form, density, electrical conductivity, and melting and boiling points are examples of
(a) liquid properties
(b) physical properties
(c) chemical properties
(d) solid properties

10. The first period contains how many elements?
(a) 2
(b) 4
(c) 6
(d) 8

11. Noble gases are also called the
    (a) actinides
    (b) chlorofluorocarbons
    (c) transuranic elements
    (d) inert gases

12. Seawater is an example of a
    (a) homogeneous solution
    (b) isotonic solution
    (c) heterogeneous solution
    (d) pasteurized solution

13. Sodium hypochlorite is commonly called
    (a) ammonia
    (b) saltpeter
    (c) bleach
    (d) rock salt

14. Sulfur is a member of the
    (a) halogens
    (b) actinides
    (c) inner transition metals
    (d) chalcogens

15. Which of the following is not a type of quark?
    (a) top
    (b) magic
    (c) up
    (d) bottom

16. Approximately what percentage of the elements are metals?
    (a) 25%
    (b) 40%
    (c) 80%
    (d) 90%

17. The actinide series of elements are
    (a) all lighter than air
    (b) radioactive
    (c) silver in color
    (d) all found in nature

18. Diamonds have what kind of crystal structure?
    (a) pentagonal
    (b) planar
    (c) elliptical
    (d) cubic

19. The molecular formula of fructose is
    (a) $C_2H_2O_6$
    (b) $C_6H_{12}O_6$
    (c) $C_6H_{10}O_2$
    (d) $C_{12}H_6O_2$

20. A molecule is the
    (a) reduction of elements to their basic structure
    (b) combination of two or more atoms held together by chemical bonds
    (c) support framework for the atom's inner sub-particles
    (d) Latin origin for the element molybdenum

21. The heavy zig-zag line to the right of the Periodic Table separates
    (a) metal salts
    (b) halogens
    (c) the actinide series
    (d) the metals and non-metals

22. A phase
    (a) is a division between metals and non-metals
    (b) something everyone goes through at one time or another
    (c) describes the number of different homogeneous materials in a sample
    (d) similar to the day and night cycle of the body

23. The halogens are all
    (a) non-metals
    (b) hydrocarbons
    (c) metals
    (d) lighter than neon

24. Americium is
    (a) a patriotic element
    (b) a member of the rare earth (lanthanide) metals
    (c) always a red metalloid
    (d) has the chemical symbol Ar

25. The atomic number of boron is
    (a) 5
    (b) 8
    (c) 10.8
    (d) 12.2

26. A Bunsen burner is
    (a) a type of chilli
    (b) a kind of sweet roll
    (c) a heat source in the laboratory
    (d) invented by Sir Alexander Bunsen

27. A cathode is another name for a
    (a) cat carrier
    (b) positively charged electrode
    (c) carbon dating device
    (d) negatively charged electrode

28. Samples that can be broken down into simpler chemicals are known as
    (a) particles
    (b) complex
    (c) compounds
    (d) cubic

29. Transuranic elements
    (a) have two-letter shorthand names
    (b) are not naturally occurring
    (c) are made in light nuclei reactions
    (d) are not highly defined

30. The total diameter of an atom is about
    (a) 100 times larger than the nucleus
    (b) 1000 times larger than the nucleus
    (c) 10,000 times larger than the nucleus
    (d) 100,000 times larger than the nucleus

31. The shorthand chemical name for protactinium is
    (a) Pt
    (b) Pr
    (c) Po
    (d) Pa

32. The periods have the following numbers of elements
    (a) 2, 8, 8, 18, 18
    (b) 2, 8, 12, 16, 18
    (c) 2, 8, 10, 12, 18
    (d) 2, 8, 10, 18, 18

33. Orbitals of the d-type come in
    (a) sets of 2
    (b) sets of 3
    (c) sets of 5
    (d) sets of 7

34. Bromine is
    (a) commonly added to toothpaste
    (b) the only liquid non-metal at room temperature
    (c) an alkaline earth metal
    (d) never found as a vapor

35. The atomic number of neptunium is
    (a) 37
    (b) 42
    (c) 93
    (d) 101

36. Who presented *On the Relation of the Properties to the Atomic Weights of the Elements?*
    (a) Lars Nilson
    (b) Antoine Lavoisier
    (c) Lothar Meyer
    (d) Dimitri Mendeleyev

37. If it is 3°C in Edinburgh, Scotland, in February, what is that in °F in Portland, Maine?
    (a) 29
    (b) 45
    (c) 63
    (d) 71

38. If the shell capacity of an orbital is $2n^2 = 8$, what is n?
    (a) $n = 1$
    (b) $n = 2$
    (c) $n = 3$
    (d) $n = 4$

39. Early humans thought rust was
    (a) red-brown dirt
    (b) a mystery
    (c) a trick of the eye
    (d) a problem with the sample

40. The lanthanides and actinides are also called the
    (a) noble gases
    (b) halogens
    (c) heavy metals
    (d) inner transition metals

41. Common table salt is made up of
    (a) lithium chloride
    (b) potassium chloride
    (c) sodium chloride
    (d) cesium chloride

42. Fer is the French name for the element
    (a) iron
    (b) phosphorus
    (c) copper
    (d) fluorine

43. The pH of milk is roughly
    (a) 5.4
    (b) 6.4
    (c) 7.4
    (d) 8.4

44. Oxygen's filled 1s orbital diagram has how many up and down arrows?
    (a) 1
    (b) 2
    (c) 3
    (d) 4

45. The Pauli exclusion principle was written by
    (a) Alfredo Pauli
    (b) Pablo Pauli
    (c) Wolfgang Pauli
    (d) Jane Pauli

46. When all orbitals of a certain sublevel have to be occupied by a single electron before pairing starts, it is known as
    (a) the Pauli exclusion principle
    (b) Schrödinger's equation
    (c) Williams' rule
    (d) Hund's rule

47. Oxidation describes
    (a) electron (+) gain
    (b) a helpful diet plan
    (c) electron (–) loss
    (d) a better spending plan

48. Nitrogen reacts with which element to form ammonia?
    (a) hydrogen
    (b) oxygen
    (c) chlorine
    (d) lithium

49. Mixtures can be divided into two types
    (a) up and down
    (b) homogeneous and heterogeneous
    (c) thick and thin
    (d) sweet and sour

50. The combining power of one element with the atoms of another is called
    (a) grouping
    (b) sensitivity
    (c) valence
    (d) nomenclature

51. This element makes up over 25% of the Earth's elements
    (a) phosphorus
    (b) potassium
    (c) calcium
    (d) silicon

52. Early alchemists related the planet Mars to which element?
    (a) iron
    (b) gold
    (c) lead
    (d) mercury

53. The first step the alchemists took to change a metal to gold was
   (a) to heat water
   (b) to get a good night's sleep
   (c) called melanosis
   (d) to check the weather

54. There are how many base units in the International System of Units (SI)?
   (a) 4
   (b) 5
   (c) 7
   (d) 9

55. A thermometer measures
   (a) color gradients
   (b) a thermos
   (c) volume
   (d) temperature

56. The shorthand chemical name for bismuth is
   (a) Bi
   (b) Bs
   (c) Bm
   (d) Bt

57. Metals that are ductile means that
   (a) they are used as a supplement in duck food
   (b) they can be pulled into wires
   (c) they can be flattened into sheets
   (d) they are used in air-conditioning ducts

58. The Nobel Prize for Chemistry is awarded to chemists
   (a) who can chew gum and jump on one foot at the same time
   (b) studying in Switzerland
   (c) with really big laboratories
   (d) who discover something new or explain something not well understood

59. If the temperature of a sample is 14°C, what is the temperature in kelvin?
   (a) 224 kelvin
   (b) 259 kelvin
   (c) 287 kelvin
   (d) 295 kelvin

60. The two electrons in each filled orbital must be
    (a) opposite in charge
    (b) really cozy
    (c) chemists
    (d) of the same charge

61. About 1% of the Earth's resources are made up of this element
    (a) silicon
    (b) manganese
    (c) phosphorus
    (d) magnesium

62. These elements are usually shiny and good conductors of heat and electricity
    (a) halogens
    (b) metals
    (c) noble gases
    (d) non-metals

63. The pH of bleach is
    (a) 5.0
    (b) 7.4
    (c) 9.2
    (d) 14.0

64. Ice is the solid form of
    (a) lead
    (b) helium
    (c) water
    (d) crystals

65. NASA Spinoffs are an example of
    (a) governmental insight
    (b) applied science
    (c) space equipment used only in outer space
    (d) pure science

66. Early alchemists used this to represent the elements
    (a) Greek letters
    (b) sign language
    (c) Arabic letters
    (d) signs of the planets

67. The SI standard unit for mass is the
    (a) gram
    (b) pound
    (c) kilogram
    (d) bushel

68. Which of the following is not a correct measurement?
    (a) grams per liter
    (b) degrees Fahrenheit
    (c) kilometers per hour
    (d) degrees per meter

69. The atomic number of hydrogen is
    (a) the same as its atomic weight
    (b) not known
    (c) only calculated when in solid form
    (d) 2

70. Elements found along the border of metals and non-metals are called
    (a) transition metals
    (b) noble gases
    (c) metalloids
    (d) rare earth metals

71. Kvicksilver is the Swedish name for
    (a) potassium
    (b) mercury
    (c) silver
    (d) tin

72. Thorium is part of which group?
    (a) halogens
    (b) lanthanides
    (c) actinides
    (d) metals

73. Nearly 2% of the Earth's elements is made up of this element
    (a) calcium
    (b) chlorine
    (c) strontium
    (d) zinc

74. Early alchemists related the planet Jupiter to which element?
(a) gold
(b) titanium
(c) silver
(d) tin

75. Which Greek philosopher called tiny particles atomos (atoms)?
(a) Plato
(b) Democritus
(c) Aristotle
(d) Lucretius

76. The distance from the Sun to Pluto is roughly
(a) 10^8 meters
(b) 10^{10} meters
(c) 10^{13} meters
(d) too far to think about

77. Hydrogen and helium are located in which period?
(a) first
(b) second
(c) third
(d) fourth

78. The SI standard unit to measure a pure substance is the
(a) chipmunk
(b) gopher
(c) mole
(d) squirrel

79. Gallium was discovered one year before
(a) copper
(b) germanium
(c) calcium
(d) selenium

80. In a binary covalent compound, phosphorus is named before which of the following elements?
(a) carbon
(b) silicon
(c) boron
(d) chlorine

81. In an ion of one atom, the oxidation number is equal to
(a) two
(b) zero
(c) the ion's charge
(d) the ion's atomic weight

82. The ion form of permanganate is
(a) MnO_4^-
(b) MgO_2^-
(c) MdH_2^-
(d) PO_4^-

83. Which of the following would not be considered toxic at this ppm level?
(a) silver (0.05 ppm in water)
(b) arsenic (0.5 ppm)
(c) diethyl ether (400 ppm)
(d) mercury (0.002 ppm in water)

84. The word radioactivity was first used by
(a) Lothar Meyer
(b) Plato
(c) Han Christian Anderson
(d) Marie Curie

85. Which element is the only one that has been given separate names for its different isotopes?
(a) hydrogen
(b) nitrogen
(c) idodine
(d) carbon

86. The two types of nuclear reactions are radioactive decay of bonds and the
(a) "billiard ball" type
(b) "tennis ball" type
(c) "golf ball" type
(d) "ping pong ball" type

87. The electronegativity value of lead is
(a) 0.6
(b) 1.5
(c) 1.7
(d) 2.2

88. Six-sided ring structures with alternating single and double bonds are called
    (a) inverse isomers
    (b) orbitals
    (c) aromatic hydrocarbons
    (d) noble gases

89. Lead forms how many kinds of ions with unique charges?
    (a) 1
    (b) 2
    (c) 3
    (d) 4

90. What is approximately the pH level of blood?
    (a) 6.5
    (b) 7.0
    (c) 7.4
    (d) 8.0

91. All oxidation numbers added together in a compound must equal
    (a) zero
    (b) one
    (c) the total number of atoms
    (d) enough to make the experiment worth doing

92. The B sub-group of elements are also called
    (a) transition elements
    (b) metals
    (c) halogens
    (d) gases

93. Which of the following is an aromatic compound?
    (a) methane
    (b) ethanol
    (c) butene
    (d) cinnemaldehyde

94. Radioactive iodine is used in the marking and treatment of
    (a) emphysema
    (b) thyroid cancer
    (c) flat feet
    (d) halitosis

95. Molecules with chiral isomers are
    (a) not superimposable
    (b) acid-base pairs
    (c) superimposable
    (d) elastomers

96. In a binary covalent compound, sulfur is named before which of the following elements?
    (a) carbon
    (b) oxygen
    (c) boron
    (d) hydrogen

97. How many elements are in the sixth period?
    (a) 18
    (b) 28
    (c) 32
    (d) 40

98. Early alchemists related the planet Saturn to which element?
    (a) boron
    (b) lead
    (c) sulfur
    (d) scandium

99. A pH meter measures the amount of
    (a) hydrogen ions in a solution
    (b) oxygen ions in a solution
    (c) nitrogen ions in a solution
    (d) sodium ions in a solution

100. The pH of most carbonated drinks is about
    (a) 3.1
    (b) 3.9
    (c) 6.2
    (d) 8.5

101. The measurement of the charge separation between each part of a molecule is called
    (a) an extra-terrestrial moment
    (b) an embarrassing moment
    (c) an tripole moment
    (d) a dipole moment

102. The size of an average water molecule is
(a) 10^{-2} meters
(b) 10^{-4} meters
(c) 10^{-8} meters
(d) 10^{-10} meters

103. Lawrencium is located in which period?
(a) third
(b) fourth
(c) sixth
(d) seventh

104. If the shell capacity of an orbital is $2n^2 = 32$, what is n?
(a) $n = 1$
(b) $n = 2$
(c) $n = 3$
(d) $n = 4$

105. Hierro is the Spanish name for
(a) lead
(b) iron
(c) mercury
(d) hydrogen

106. *Covalent* bonds occur between two
(a) bases
(b) metals
(c) non-metals
(d) a metal and a non-metal

107. If the temperature of a sample is 273 kelvin, what is the temperature in °C?
(a) 0
(b) 10
(c) 32
(d) 73

108. Which element melts in your hand like a chocolate bar?
(a) krypton
(b) tantalum
(c) strontium
(d) cesium

109. Carbon single-walled, nano-scale tubular structures are called
(a) bucky balls
(b) rare earth metals
(c) nanotubes
(d) chalogens

110. One of the first acids a beginning chemistry student uses in the lab is
(a) hydrochloric acid
(b) cyanoic acid
(c) propionic acid
(d) barbaric acid

111. Proteins give hard structural strength in all but which of the following?
(a) ankle bone
(b) crab shell
(c) spinal column
(d) earthworm

112. The electronegativity value of hafnium is
(a) 0.7
(b) 1.3
(c) 1.8
(d) 2.2

113. Lutetium is part of which group?
(a) halogens
(b) lanthanides
(c) actinides
(d) metals

114. What is the approximate pH of stomach acid?
(a) 2.0
(b) 4.9
(c) 6.2
(d) 7.5

115. Many people with milk sensitivities have trouble with the milk sugar
(a) fructose
(b) lactose
(c) sucrose
(d) glucose

116. If a solution measured 5.8 on the pH meter, it would be
    (a) a radioactive solution
    (b) a noble gas
    (c) a base
    (d) an acid

117. Litmus paper changes from blue to red when a solution tested is
    (a) a blue dye
    (b) an acid
    (c) a red dye
    (d) a base

118. Bases react with acids to form
    (a) dark chocolate and white chocolate
    (b) acetone and water
    (c) water and salt
    (d) propane and butane

119. An uncombined element has an oxidation number of
    (a) 0
    (b) 1
    (c) 2
    (d) 3

120. In a binary covalent compound, iodine is named before which of the following elements?
    (a) bromine
    (b) silicon
    (c) sulfur
    (d) nitrogen

121. Early alchemists related the planet Venus to which element?
    (a) iodine
    (b) copper
    (c) potassium
    (d) sodium

122. The SI standard unit for length is the
    (a) centimeter
    (b) decimeter
    (c) meter
    (d) kilometer

123. Cesium is located in which period?
(a) second
(b) third
(c) fifth
(d) sixth

124. The pH of vinegar is
(a) 3.0
(b) 3.6
(c) 4.2
(d) 5.0

125. Nickel forms how many kinds of ions with unique charges?
(a) 1
(b) 2
(c) 3
(d) 4

126. Which of the following is an aromatic compound?
(a) propanol
(b) isobutane
(c) vanillin
(d) ethyne

127. Which of the following is the most widely produced acid in the United States?
(a) sulfuric acid
(b) nitric acid
(c) boric acid
(d) malic acid

128. If a solution measured 7.7 on the pH meter, it would be
(a) a halogen
(b) a macromolecule
(c) a base
(d) an acid

129. If a solution measured 8.9 on the pH meter, it would be
(a) a brittle solid
(b) an inert gas
(c) a base
(d) an acid

130. When oxygen is part of a compound, the oxidation number is −2 except for
(a) salts
(b) peroxides
(c) hydrocarbons
(d) nanomolecules

131. If the temperature of a sample is 343 kelvin, what is the temperature in °C?
(a) 50°C
(b) 70°C
(c) 90°C
(d) 110°C

132. Early alchemists related the Sun to which element?
(a) gold
(b) silver
(c) copper
(d) iron

133. In 1923, two chemists, Johannes Brönsted and Thomas Lowry, described
(a) mesons and hadrons
(b) acids and bases in the scientific literature
(c) the Periodic Table
(d) how to construct nanotubes

134. Litmus paper changes from red to blue when a solution tested is
(a) a blue dye
(b) an acid
(c) a red dye
(d) a base

135. Which of the compounds below do most people learn about first in chemistry?
(a) lava
(b) vanillin
(c) water
(d) ozone

136. Reduction is the chemical name for
(a) when a balloon loses air and shrinks
(b) a decrease in oxidation number

(c) the loss of energy from a gas
(d) the process of writing a name on a grain of rice

137. Oxygen has an oxidation number of
(a) −1
(b) −2
(c) −3
(d) −4

138. Helium is how many times denser than hydrogen?
(a) 2 times
(b) 3 times
(c) 4 times
(d) helium is the same density as hydrogen

139. The invention of the Gutenberg printing press in 1452 helped spread which theory?
(a) Theory of Relativity
(b) Nano-technology Theory
(c) Conservation of Energy Theory
(d) Atomic Theory

140. In 1995, three chemists, Molina, Rowland and Crutzen became concerned about damage to the
(a) polar ice caps
(b) Golden Gate Bridge
(c) ozone layer
(d) Canadian Rockies

141. Chemically, simple sugars are also known as
(a) something to avoid when on a diet
(b) monosaccharides
(c) the only cause of dental cavities
(d) disaccharides

142. Antimony forms how many kinds of ions with unique charges?
(a) 1
(b) 2
(c) 3
(d) 4

143. Argon is located in which period?
(a) first
(b) second

(c) third
(d) fourth

144. Early alchemists related the Moon to which element?
(a) manganese
(b) vanadium
(c) chromium
(d) silver

145. The pH of seawater is roughly
(a) 5.2
(b) 6.1
(c) 7.7
(d) 8.9

146. Dipole moments are measured in
(a) coulomb-meters
(b) flashes per minute
(c) grams per inch
(d) kilometer-hours

147. A complex carbohydrate found in plant cell walls is called
(a) chitin
(b) cellulose
(c) caramel
(d) kilometer-hours

148. Bly is the Swedish name for the element
(a) boron
(b) beryllium
(c) lead
(d) bromine

149. The size of an average earthworm is about
(a) 10 meters
(b) 10^2 meters
(c) 10^{-2} meters
(d) 10^{-4} meters

150. Most nanomolecules that transfer electrical charges are
(a) the size of soccer balls
(b) found only in underwater grottos
(c) handled with rubber gloves to prevent shock
(d) carbon-based polymers

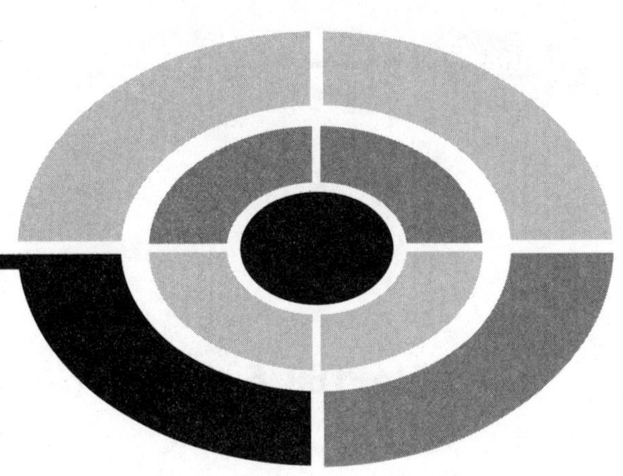

Answers to Chapter Quiz Questions

CHAPTER 1

1. B.	2. D	3. D	4. C	5. A
6. C	7. C	8. B	9. D	10. A

CHAPTER 2

1. C	2. D	3. C	4. A	5. B
6. C	7. D	8. B	9. A	10. B

CHAPTER 3

1. C	2. D	3. B	4. A	5. D
6. C	7. A	8. B	9. C	10. D

CHAPTER 4

1. C	2. B	3. D	4. D	5. C
6. A	7. B	8. C	9. A	10. D

CHAPTER 5

1. B	2. A	3. D	4. A	5. B
6. A	7. D	8. B	9. C	10. A

CHAPTER 6

1. B	2. C	3. D	4. C	5. A
6. B	7. D	8. C	9. D	10. B

CHAPTER 7

1. C	2. A	3. D	4. C	5. B
6. A	7. D	8. D	9. C	10. B

CHAPTER 8

1. B	2. A	3. B	4. D	5. D
6. A	7. B	8. D	9. C	10. C

CHAPTER 9

1. C	2. C	3. A	4. D	5. B
6. B	7. C	8. D	9. A	10. B

CHAPTER 10

1. C	2. D	3. A	4. C	5. D
6. B	7. A	8. B	9. D	10. C

CHAPTER 11

1. D	2. B	3. C	4. B	5. D
6. B	7. B	8. D	9. A	10. D

CHAPTER 12

1. C	2. B	3. D	4. A	5. C
6. D	7. D	8. B	9. A	10. D

CHAPTER 13

1. C	2. A	3. C	4. D	5. B
6. A	7. D	8. D	9. C	10. B

CHAPTER 14

1. A	2. D	3. B	4. A	5. D
6. C	7. B	8. D	9. C	10. D

CHAPTER 15

1. B	2. C	3. A	4. B	5. D
6. B	7. C	8. A	9. D	10. B

CHAPTER 16

1. B	2. D	3. C	4. D	5. B
6. A	7. D	8. C	9. A	10. A

CHAPTER 17

1. D	2. B	3. C	4. D	5. A
6. B	7. A	8. D	9. C	10. B

CHAPTER 18

1. C	2. D	3. B	4. A	5. B
6. D	7. C	8. B	9. D	10. A

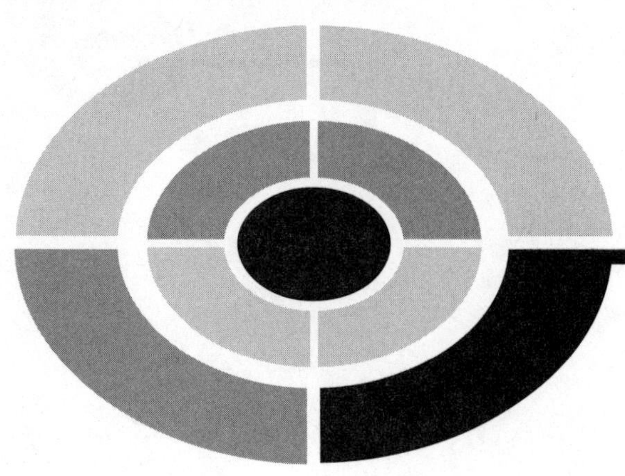

Test: Part One Answers

1. B	2. D	3. C	4. D	5. B
6. A	7. C	8. C	9. D	10. B
11. B	12. C	13. D	14. A	15. B
16. B	17. C	18. D	19. D	20. A
21. D	22. A	23. C	24. D	25. B
26. A	27. C	28. B	29. D	30. A
31. B	32. D	33. C	34. B	35. A
36. A	37. C	38. A	39. D	40. D

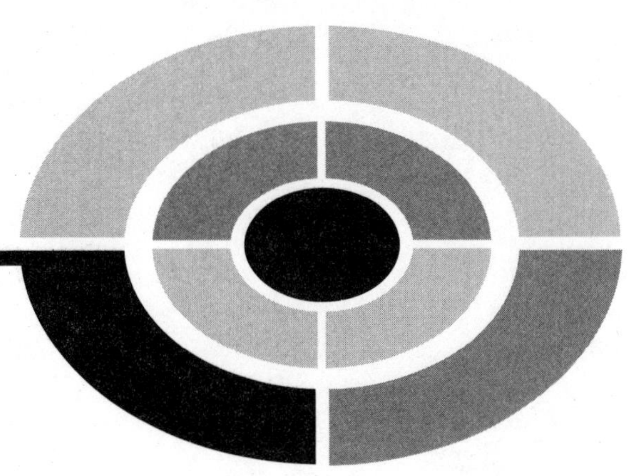

Test: Part Two Answers

1. B	2. C	3. D	4. B	5. D
6. A	7. C	8. B	9. D	10. C
11. C	12. B	13. A	14. C	15. D
16. B	17. C	18. B	19. C	20. A
21. D	22. B	23. C	24. D	25. B
26. A	27. D	28. B	29. B	30. C
31. D	32. B	33. A	34. D	35. A
36. B	37. C	38. B	39. C	40. D

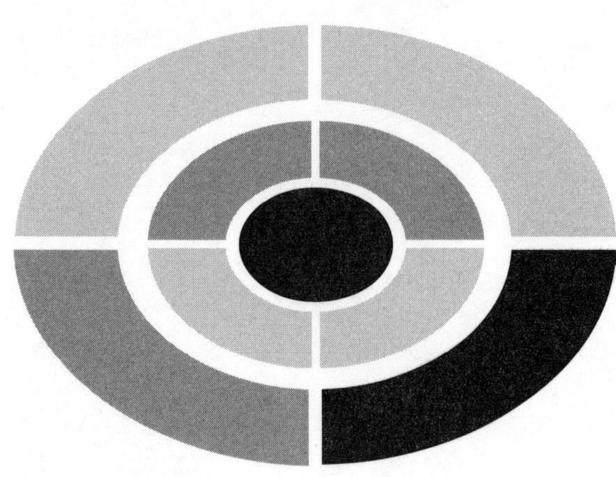

Test: Part Three Answers

1. A	2. D	3. B	4. D	5. B
6. C	7. B	8. C	9. B	10. B
11. B	12. A	13. D	14. A	15. C
16. C	17. B	18. B	19. A	20. B
21. B	22. D	23. D	24. B	25. D
26. C	27. B	28. C	29. A	30. B
31. C	32. D	33. C	34. D	35. B
36. A	37. C	38. D	39. B	40. C

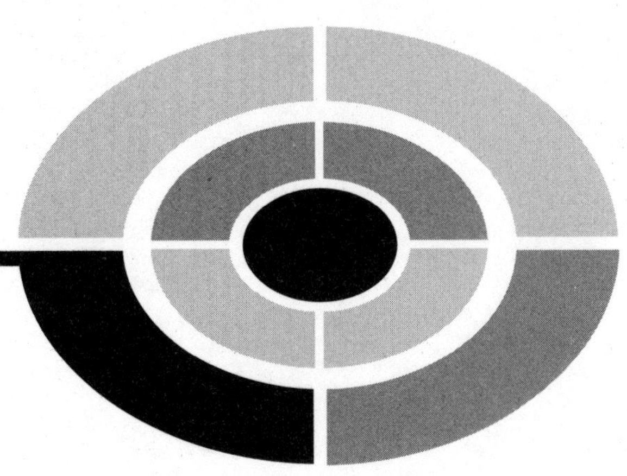

Test: Part Four Answers

1. C	2. D	3. B	4. B	5. D
6. C	7. A	8. B	9. B	10. A
11. C	12. D	13. B	14. A	15. B
16. C	17. C	18. A	19. D	20. B
21. A	22. C	23. A	24. B	25. D
26. C	27. B	28. D	29. C	30. D
31. A	32. B	33. B	34. C	35. D
36. C	37. C	38. A	39. B	40. B

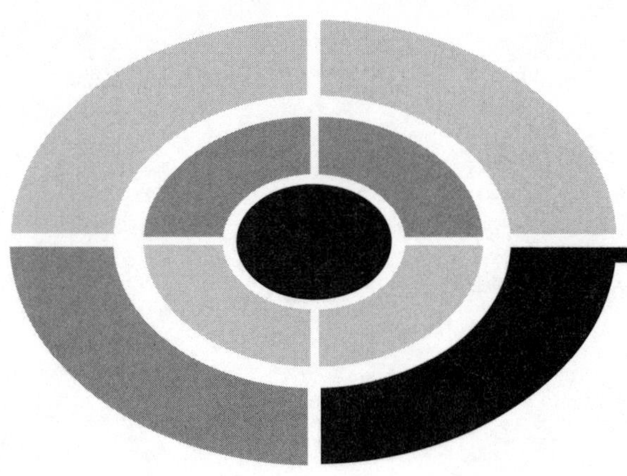

Final Exam Answers

1. D	2. A	3. D	4. D	5. C
6. C	7. B	8. C	9. B	10. A
11. D	12. A	13. C	14. D	15. B
16. C	17. B	18. D	19. B	20. B
21. D	22. C	23. A	24. B	25. A
26. C	27. D	28. C	29. B	30. C
31. D	32. A	33. C	34. B	35. C
36. D	37. C	38. B	39. B	40. D
41. C	42. A	43. B	44. B	45. C
46. D	47. C	48. A	49. B	50. C
51. D	52. A	53. C	54. C	55. D
56. A	57. B	58. D	59. C	60. A
61. C	62. B	63. D	64. C	65. B
66. D	67. C	68. D	69. A	70. C
71. B	72. C	73. A	74. D	75. B
76. C	77. A	78. C	79. B	80. D
81. C	82. A	83. B	84. D	85. A
86. A	87. D	88. C	89. B	90. C
91. A	92. A	93. D	94. B	95. A
96. B	97. C	98. B	99. A	100. B

101. D	102. C	103. D	104. D	105. B
106. C	107. A	108. D	109. C	110. A
111. D	112. B	113. B	114. A	115. B
116. D	117. B	118. C	119. A	120. A
121. B	122. C	123. D	124. A	125. B
126. C	127. A	128. C	129. C	130. B
131. B	132. A	133. B	134. D	135. C
136. B	137. B	138. A	139. D	140. C
141. B	142. B	143. C	144. D	145. C
146. A	147. B	148. C	149. C	150. D

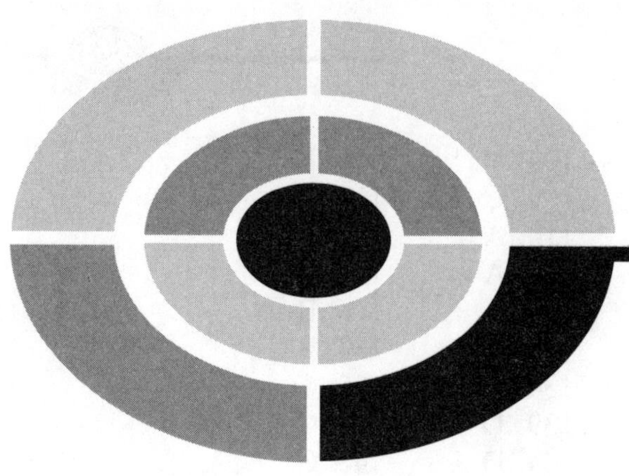

References

Akins, Peter William. *The Periodic Kingdom: A Journey into the Land of the Chemical Elements*. New York: Basic Books, 1995.

Akins, Peter William. *Molecules*. New York: Scientific American Library Series No. 21, 1987.

Ball, Philip. *Stories of the Invisible: A Guided Tour of Molecules*. New York: Oxford University Press, 2001.

Ebbing, Darrell D. *General Chemistry*. Boston: Houghton Mifflin Company, 1993.

Emsley, John. *Molecules at an Exhibition: Portraits of Intriguing Materials in Everyday Life*. New York: Oxford University Press, 1998.

Fernando, Diana. *Alchemy: An Illustrated A to Z*. New York: A Blandford Book, 1998.

Johnson, Moira. *The Facts on File Chemistry Handbook*. New York: Checkmark Books, Diagram Visual Information Ltd., 2001.

Malone, Leo J. *Basic Concepts of Chemistry*. New York: John Wiley & Sons, Inc., 1976.

Smiley, Robert A. and Jackson, Harold L. *Chemistry and the Chemical Industry: A Practical Guide for Non-Chemists*. New York: CRC Press, LLC, 2002.

Solomons, T. W. Graham. *Organic Chemistry*. New York: John Wiley & Sons, Inc., 1976.

Strathern, Paul. *Mendeleyev's Dream*. New York: Thomas Dunne Books, 2000.

Stwertka, Albert. *A Guide to the Elements*. New York: Oxford University Press, 1996.

Waites, Gillian and Harrison, Percy. *The Cassell Dictionary of Chemistry*. London: Cassell, 1999.

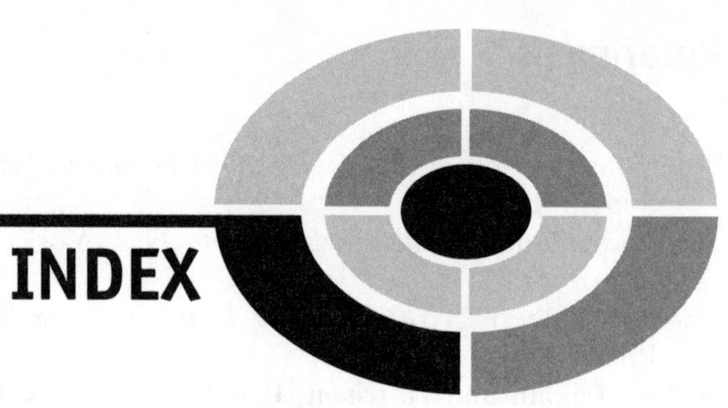

INDEX

ABOUT THE AUTHOR

Linda Williams is a nonfiction writer with a specialty in science and medicine. A resident of Houston, Texas, Linda's work has ranged from biochemistry and microbiology to genetics and human enzyme research. She has worked as a technical writer and lead scientist for NASA and McDonnell Douglas, and has served as science speaker for the Medical Sciences Division at NASA-Johnson Space Center in Houston.